THE WARNING

ACCIDENT AT THREE MILE ISLAND

THE WARNING

ACCIDENT AT THREE MILE ISLAND

MIKE GRAY & IRA ROSEN

Contemporary Books, Inc.
Chicago

Library of Congress Cataloging in Publication Data

Gray, Mike.
  The warning.

  Reprint. Originally published: 1st ed. New York:
Norton, © 1982.
  Includes index.
  1. Atomic power-plants—Pennsylvania—Accidents.
2. Three Mile Island Nuclear Power Plant (Pa.)
I. Rosen, Ira. II. Title.
TK1345.H37G7 1983   363.1'79   82-22167
ISBN 0-8092-5547-2 (pbk.)

Book design by Jacques Chazaud

Published by Contemporary Books, Inc.
180 North Michigan Avenue, Chicago, Illinois 60601
Manufactured in the United States of America
International Standard Book Number: 0-8092-5547-2

Published simultaneously in Canada by
Beaverbooks, Ltd.
150 Lesmill Road
Don Mills, Ontario M3B 2T5
Canada

This edition published by arrangement
with W. W. Norton & Company, Inc.

To Frank Wills, the Watergate
security guard who saw something
fishy and called the cops.

And to Jim Creswell, a former Nuclear
Regulatory Commission inspector
who tried to do the same thing.

AUTHORS' NOTE

Shortly after eight o'clock on the morning of March 28, 1979, word began to filter in to NRC Headquarters in Bethesda, Maryland, that an unplanned event had occurred some hours earlier at Three Mile Island. A PR man from the regional office was actually the first to get through, but other calls quickly followed and it was soon clear that the incident was still underway. At 8:10, the executive director of the agency ordered the staff to activate the Emergency Response Center. Moments later, as required by federal regulation, a twenty-six-channel tape machine began recording all incoming and outgoing telephone calls.

A few hours later, when Chairman Joe Hendrie voted to close the NRC board meetings to public and press, the chairman and his four colleagues were joined by a gentleman named Beckwitt—the "sunshine officer"—who followed them from room to room and building to building over the next several days, tape-recording everything they said. Mr. Beckwitt's presence was mandated by a statute that requires the head of any federal agency to provide a transcript of closed meetings. The original intent of the law was to keep the bureaucrats of the post-Nixon era a little more honest in their dealings; no one ever imagined a situation like this.

The tapes are, by definition, public property. The result of these artifacts of democracy is a stack of paper several feet high. Upon close scrutiny, it tells an interesting story, but it is only the tip of the iceberg.

Because of the importance of the issues surrounding the accident, President Carter immediately announced the appointment of a blue-ribbon investigative body to be headed by John Kemeny, the president of Dartmouth College. Carter's commission was actually third in line. Two other investigations were already underway; a fourth and a fifth would follow, making the incident at Three Mile Island one of the most thoroughly examined events in U. S. history. In the course of these five separate inquiries, everyone even remotely connected with the plant—builders, designers, architects, inspectors, operators, and owners—was questioned under oath, some as many as eight times. The 50,000 pages of depositions left in the wake of these investigations, along with the tapes, transcripts, and engineering data, form the basis for this book.

The authors met in Harrisburg on the third day of the accident. Ira Rosen, a young New York reporter, now a CBS "60 Minutes" producer, had taken a tip from the movie *The China Syndrome* and stationed himself at a bar near the plant, where he was able to talk to the control-room operators as they came off shift; unfortunately, since he is not an engineer, Rosen needed help in understanding the technical details of what they were saying. Mike Gray is an engineer. He had spent five years researching nuclear accidents before writing the original screenplay for *The China Syndrome,* and was able to clarify the interviews that Rosen extracted. A few days later, they agreed to collaborate on this book. Over the next two years they conducted their own extensive interviews of all the principal figures in the event, from plant-maintenance personnel to the governor of Pennsylvania, and this yielded another 200 hours of transcripts.

This mountain of facts was entered into the computer memory of a WANG System 5 Word Processor and reorganized for instant access. From this foundation, after thirty months of sifting, an account of the accident and the events leading up to it was reconstructed. The finished draft was then checked for accuracy by several of the main characters in the story. What follows is, to the best ability of everyone involved, a frightening but unbiased narrative of the country's closest brush with nuclear catastrophe.

ACKNOWLEDGMENTS

There have been many people who helped us with this book, but we have been fortunate beneficiaries for having the aid of Dr. Henry Kendall, Maryanne Vollers, Peter Bradford, Henry Myers, James Dennett, MHB consultants, Robert Pollard, Paul Goldberg, Carol Gray, Leo and Ethel Rosen, and the Kinney family of Middletown, who took us in and gave us their trailer to work and live in; also, of the kind people at the Tabard Inn, where parts of this book were written. And we would also like to thank Babcock & Wilcox and Metropolitan Edison, who opened their control rooms to us.

Finally, for indispensable guidance given with rare grace and *much* patience, we are forever grateful to our editor, Star Lawrence, and his assistant, Kathleen Anderson.

Mike Gray
Ira Rosen

New York City
December 1981

ILLUSTRATIONS

Drawings by Mike Gray

Photographs by P. Michael O'Sullivan

Engineering is an art form that makes use of scientific principles, and this marriage confuses a lot of people. We tend to think of engineering itself as a science, but it is nothing more than advanced carpentry. The practitioners learn by doing, and the craft is constantly in a state of evolution. When the Three Little Pigs built their houses, they all had acceptable designs—acceptable to them—but two of the piggies underestimated the wind loads.

THE
WARNING
ACCIDENT AT THREE MILE ISLAND

1

December 1968

From the boardroom window, Jim Neely looks out across New York Harbor, and his thoughts are as dark as the winter sky over Jersey. Neely has been around several large construction projects but never one like this. This deal is a winner, possibly a record for extortion.

Across the Hudson, labor corruption is a tradition as hallowed as baseball. Neely had heard about construction in New Jersey where kickbacks were paid to union officials. It was all very out-front; somebody's cousin shows up for work and gets paid $25,000 a year to do nothing. Complain about it, and you get the shit kicked out of you. But this time the boys may have stepped over the line. It's one thing to stick up the construction boss at the plant site; it's something else to try to rough up corporate headquarters.

When the company Neely worked for, Jersey Central Power and Light, announced plans to build Oyster Creek Unit Two, the mob must have thought it was Christmas in July. One of the boys went to the president of the company and told him he would guarantee industrial peace for the duration in exchange for 1 percent of the construction price. Since the second nuclear plant at Oyster Creek is expected to run $700,000,000, this guy would pocket $7 million.

No question he means business. To underline this point,

somebody dropped a wrench in a crane hoist while they were lowering the 700-ton reactor vessel in Unit One; if it hadn't been spotted in time, the crew might have dropped the reactor.

But the mobster, mean and powerful as he may be, has an inflated idea of his importance. His adversaries are much bigger than he realizes. Jersey Central Power and Light is only one of the chips on the table, one piece of the holding company known as General Public Utilities. GPU has subsidiaries outside of New Jersey, and right now their operations in Pennsylvania are looking more and more attractive.

This meeting has been called at corporate headquarters to see if they can simply move the plant out of New Jersey. It's too late to dump the project; most of the design work has been completed; they're already down $20 million.

Neely takes his chair and looks around the table. Along with the technical v.p.'s from every division in the company, there are people from Burns and Roe, the plant architect-engineers, and from Babcock & Wilcox, the designers of the reactor. The man with the flip charts is Lou Roddis, head of the GPU nuclear group, who's been studying the feasibility of the move.

Roddis and his team have considered all the factors, and they believe the plant can be relocated; they've come up with a couple of alternate sites. At one location, it could be built alongside a similar reactor already under construction. It would require some major changes—for one thing, the reactor building would have to be rotated ninety degrees on the blueprints to mate with the existing structure. This would change the relationship of all the piping, but it's a problem they can handle.

Management is less concerned about the hardware than about the union and the government. Getting an operating license from the Nuclear Regulatory Commission has become a ten-year endurance race; fall out and you go back to the starting gate. But Roddis says if they act quickly, it might be possible to maintain the existing schedule. At worst, he estimates, they will slip eight months. Since this is a time frame that the company, the banks, and the stockholders can live with, the issue is settled then and there.

But Jim Neely feels more than a little uncomfortable. There's a hell of a lot of work to be done in order to maintain that schedule. A lot of the hardware that was designed for use at Oyster Creek will have to be extensively modified. But like everybody else at the table, Neely would rather deal with engineering problems than with corrupt union leaders. The guy who wanted 1 percent isn't the only extortionist in New Jersey; he was just the first one to show up.

Working at a killing pace over the next ninety days, the combined engineering staffs of these several corporations manage to completely rewrite the entire design study for the plant. On March 10, 1969, a company courier arrives in Washington with an amendment to the GPU petition for license. Included is a complete revision of the *Preliminary Safety Analysis Report,* a five-volume study running more than a thousand pages. The amendment requests that Jersey Central Power and Light be allowed to transfer ownership of the plant to Metropolitan Edison of Pennsylvania, and that the plant be relocated from Oyster Creek in New Jersey to Three Mile Island on the Susquehanna south of Harrisburg.

2

Toledo, September 24, 1977

"What was *that*, for Christ's sake?"

From the shift supervisor's office, it sounds like the gates of hell. Mike Derivan rises from his desk and makes his way to the control room. The rumble is coming from the turbine building. For sure, the safety valves have lifted.

The long sweep of the instrument console is awash with blinking lights, and the room is filled with sounds of alarm as half a dozen operators try to make sense out of the incoming signals. Hunched over the center desk, Roger Brubaker catches a glimpse of Derivan as he enters. He shouts above the din, "Pressurizer level's going straight up! Two hundred ninety inches. I'm gonna trip the reactor."

"Go ahead!"

Tripping the reactor is no more serious than blowing a fuse; it happens all the time. But each time it's a little different, and there is something definitely different about this one.

"What happened?"

"Lost level in the number-two steam generator. Pumps or something."

On the low console before them are the master instruments that show the general health of the exotic machine they supervise. They watch in amazement as the pressure in the reactor drops several hundred pounds in less than a minute.

Out on the main deck of the turbine building, Assistant Station Superintendent Terry Murray doesn't need a weathervane to know which way the wind is blowing. With the first roar of venting steam, he begins making his way to the control room. He arrives at about the same time as the operations engineer, Bob Zemenski. Though it is the dark of night, it's not unusual for these Toledo Edison executives to still be at the plant. Davis-Besse nuclear station was licensed to operate less than six months earlier; the reactor and its crew are just getting to know each other. Everybody has been putting in sixteen-hour days trying to work out the kinks. Looks like another long one tonight.

Ranging around the instrument panels that line the control room, the operators are leafing through emergency manuals, verifying valve lineups, and shouting numbers. At the center console, Derivan and Brubaker are trying to figure out what the hell is going on. So far nothing fits. It is pressure that keeps the water in the core from boiling. And as the pressure continues to drop with each wink of the meter, the reactor takes control to protect itself.

"We got HPI"

"Verify that."

High-pressure injection: the first line of defense against nuclear disaster. Down below, in the concrete vaults of the auxiliary building, a trio of five-foot high pumps gets a call to service from the computer. In seconds the screaming impellers are pushing water into the core at a pressure of nearly a ton to the square inch.

On the console beside Derivan, the alarm printer rattles out line after line of numbers, but he ignores it. Events are moving too quickly now, and the printer is running far behind. From everywhere within the great plant, from the miles of pipe and pumps and valves, the rows of tanks and pressure vessels, the generators and turbines, comes a flow of information to the control room. If, at any point in this labyrinth, the pressure or temperature or level or flow is not what it should be, an alarm sounds and the operators must acknowledge it before the bell will stop. But the alarms are coming too fast: one after the other, a wild cacophony of lights and bells, stunning to the brain.

Derivan is no neophyte in this business. Navy trained, he rose in grade to be engine-room supervisor. He's been to lab courses at the University of Michigan. He spent three months on a simulator at the Babcock & Wilcox factory in Lynchburg, Virginia. But he has never seen anything like this.

The water level is rising again. After a quick discussion they agree that the system must be overfilling because of the emergency pumps. "We've got the pressurizer level under control; I'm going to turn off the high-pressure-injection pumps!"

"Go ahead!"

As the water level starts to descend, the tension in the control room eases with it. Whatever it is, everything seems ——What the hell is this? Water level is . . . increasing?

Suddenly Derivan feels a slight tightening across his chest. This is not going to have a pleasant resolution. At best there will be damage to the plant. At worst, there could be a disaster so staggering that it is officially never discussed. Somehow Derivan has to get a grip on this thing, get it stabilized. The vast plant stretching beyond him, acre on acre of concrete and steel, a precision machine of gargantuan scale with power enough for a whole city, is out of control.

Grasping at straws now, somebody suggests that the main cooling pumps should be cut off. The idea is that this would reduce heat input to the system, since the pumps are so large they generate a great deal of heat all by themselves. The concept is insane under the circumstances, but their confusion is complete. They decide to compromise by cutting off half the pumps.

For about thirty seconds, it looks like a good idea. The water level is still rising, but much more slowly. Then the needle seems to turn a corner, and suddenly the level shoots upward.

"Two hundred seventy."

"Two hundred eighty."

"Two hundred ninety."

"Three hundred!"

And off the scale. One thing they drill into you in the training program: the system must never be completely full of water. If you don't have a bubble of steam at the top to act as a shock absorber, any sudden jolt could fracture the pipes. The tension at the center console is mind numbing, but the bells are relentless. Of the six hundred alarms lining the walls, over half are flashing. And still they come.

"Number-two steam generator! Low level!"

"It's boiling dry!"

"Look at this. The auxiliary pump's screwed up."

"Take it manual."

They might be able to grasp the situation if they could see all the facts in a single set of instruments, but the gauges they should be watching are spread all over the room. There's no way to visualize the whole system at once. But somehow in the cascade of flashing lights, Derivan's eye manages to catch an essential signal.

CONTAINMENT PRESSURE HI

High pressure in the containment building? Derivan sprints around the console to check the back panel for himself. On two of the meters, the needle is just lifting off the peg, but the third already shows three pounds pressure per square inch. At last everything fits. The pieces of mosaic resolve themselves into a clear picture. Derivan shouts, "Shut the block valve!"

He's got it now. One of the relief valves must have jammed open right in the first few seconds. The pressure's been leaking through that goddamn valve. In the twenty minutes since this thing started they have been whipsawed by the sudden shifts, and they are drenched with exhaustion. But the next sudden change is for the better. At twenty-six minutes, the water level comes back on scale.

There is damage down below, who knows how much. But the excitement is over. Outside, the sun is rising over Lake Erie. The empty horizon silhouettes an ore boat bound for the mills of Toledo. And to the east, the city of Cleveland sleeps on, unaware of its brush with destiny.

When October rolls over the Shenandoah Mountains into Lynchburg, the aging Virginia mill town is washed by the waves of color that attracted the eye of Thomas Jefferson and three other United States presidents who lived among these hills. But Joe Kelly is glad to be home for more reasons than the view along the Rappahannock. The two days in Toledo were slightly unnerving.

Kelly's boss—and a number of other Babcock & Wilcox engineers—are curious about the incident at Davis-Besse: sufficiently curious that they sent him up a couple of days afterward to look into the details. This is unusual, but the event seemed a little complicated.

At thirty-six, Joe Kelly looks like a recruitment poster for the engineering profession. His mustache, his sideburns, his hair—all are long enough to suggest youthful ideas, but not so long as to distress the century-old manufacturing company he works for. Babcock & Wilcox is a conservative organization, and Kelly is the picture of the upwardly mobile engineer.

Also, he is conscientious, and there are a number of things about the incident at Davis-Besse that disturb him. To begin with, the operators managed to blow 11,000 gallons of coolant out of the main loop before they figured out that the valve was stuck. But the unnerving thing is that they managed to get the water boiling around the core. There is a possibility the fuel is damaged.

On the plane with Kelly is Fred Faist, an experienced young field engineer who has been with the company since he got out of college. Faist is the Babcock & Wilcox man in Toledo; he was one of the first on the scene after the accident.

There is a sense of urgency about all this. The plant is crawling with government inspectors. There was a big meeting at the Davis-Besse plant while Kelly was there, and there must have been twenty people from the Nuclear Regulatory Commission. These guys want answers, and Toledo Edison will be out on a limb until the plant's designers can figure out what went wrong. When the company plane touches down at Lynchburg Airport, a company car is waiting.

The low brick building on Old Forest Road is headquarters for the Nuclear Power Generation Division of Babcock & Wilcox. Nothing is actually built here. Fabrication of the 700-ton pressure vessels, the 400-ton boilers—all that blistering muscle work where a man can get his arm ripped off—all that goes on elsewhere. This building houses the thinkers. Here in this building a man can draw lines on paper, and somewhere in Gary or Pittsburgh, hammer mills and break presses will form giant sculpture to the nearest thousandth of an inch.

In the age of heavy technology, it is the engineer who wields day-to-day power. Just as the stonecutters of medieval Europe seemed endowed with mystical powers as they raised cathedrals beyond understanding, the engineers seem removed, separated—possessors of special knowledge, speaking a special language.

In *Fire on the Moon*, Norman Mailer finally recognizes that the NASA engineers eat at the vending machines not because they don't have anybody to go to lunch with, but because they don't want to break their train of thought. At Babcock & Wilcox headquarters, the vending machines are a featured item. Halfway down the main corridor is a spacious alcove where you can buy anything from chicken soup to Bubble Up without breaking your train of thought.

Down the wide hall beyond the coffee machine is the entrance to the training department. On the right is the simulator—a computer-powered duplicate of an actual control room—where the company trains people who operate and manage nuclear-power plants. To the left is a classroom big enough to hold a couple of dozen students, but this morning it is crowded with executives.

Bruce Karrasch, head of plant integration, is here. Also Bert Dunn, chief analyst for the emergency core-cooling section; Bob Jones from plant design, and Joe Lauer from licensing. About thirty people altogether. A big deal for Kelly, and it gets bigger a few minutes into his presentation, when the Old Man himself squeezes into the room.

Vice-President John H. MacMillan, after forty years in the industry, has arrived at a point quite near the pinnacle of his profession. He is the number-one engineer in the newest, most exotic division of one of the oldest companies in America. Babcock & Wilcox is a relative newcomer to nuclear engineering. In a field dominated by General Electric and Westinghouse, B & W has sold only eight of the seventy nuclear-power plants operating in the United States. But the prospects are bright. A dozen more are on order or in the talking stage.

In a business where your cheapest item costs several million dollars, your reputation is always on the line. Since the

1860s, Babcock & Wilcox has built high-pressure boilers for everything from jute mills to battleships, and their reputation is as solid as casement steel. It is John MacMillan's responsibility to keep it that way. He has come down here to get a personal feel for the situation in Toledo. The seats are taken, so he joins several others standing against the back wall. Fred Faist is talking.

"I don't recall exactly what time it was. I received a phone call from the plant superintendent, Jack Evans, who indicated that the plant had tripped, the power-relief valve had stuck open, the rupture disc on the quench tank had ruptured. There had been some damage internally, and the plant was presently in a cool-down." The accident began when a stray computer signal cut off one of the pumps feeding water to the boilers. One thing led quickly to another.

For close to an hour, Kelly and Faist reconstruct the event as various engineering-department heads fire questions at them. Faist himself is particularly concerned about the possibility of fuel damage. He's an expert on core-heat problems, and he knows that steam in the reactor is bad news. Once you get steam bubbles in the water, it doesn't carry away the heat as effectively. It's possible some of the slender fuel pins have cracked.

But it turns out their fears are groundless. Bob Jones, one of the plant-design engineers, examines the temperature tracings that Kelly brought back from Toledo. Everyone is greatly relieved when Jones tells them they can forget about core damage; the boiling was stopped in plenty of time.

Don Montgomery, one of the men on MacMillan's personal staff, wants the details on the stuck relief valve. This device sits at the very top of the main cooling system and has a tricky job. In any kind of boiler or steam system, safety valves are required by law. This is a carryover from the era of Casey Jones, when an overzealous stoker or a sloppy riveting job sometimes blew the interested onlookers to kingdom come.

Live steam is incredibly powerful; to keep it contained, an engineering code was developed to insure that the weak link would not be in the welding or the piping, but rather in a fail-safe relief valve set to blow well before the pipes or welds are in danger of ripping open at the seams. These valves are known as the "code safeties." When they blow, they create a fearful mess. In a nuclear plant, the escaping steam contains chemicals that turn to acid; when this sprays all over the high-priced hardware in the reactor building, it can make an accountant clutch his heart.

So a special relief valve, leading to a small tank, was installed right beside the code safeties; its job is to take care of little bumps in pressure and leave the main valves undisturbed. This device—called the "power relief valve"—is designed to bleed off a little steam, then close. From a design standpoint this is very demanding. The valve must open at 2,200 pounds per square inch, then reclose when the pressure drops to 1,800 pounds. Since the throat of the valve is about four square inches, it takes over three and a half tons of force to close it.

It doesn't work all the time—in fact, it has a history of not working—but when it works it saves a lot of time and money. At Davis-Besse, the power relief valve worked too well. The tracings show that it opened and closed nine times in a few seconds, and after hammering itself to pieces, failed in the open position. Ironically, the valve itself was not at fault. Kelly and Faist discovered that an electrical relay was missing from the control circuit.

These relays—electronic switches—are small, plug-in devices, and all that Kelly and the others found was the empty socket. The relay was gone. Somebody raises the question of sabotage. Kelly dismisses that. It turns out that the missing relay is a common item in the Davis-Besse control circuits. This particular model is in short supply. It was probably just borrowed.

A group is selected to fly back to Davis-Besse the following day to assist the customer in recovery operations and to make an assessment of the plant's condition that will satisfy the government. Kelly will not be with them. He's done his job, and he has no interest in going back to Toledo. But as the meeting winds down and breaks into groups of specialists, one of the chief engineers approaches Kelly with a chilling revelation.

Bert Dunn is responsible for the design of the emergency core-cooling system. It was his group that conceived and tested the high-pressure-injection system, the core flood tanks, and the dozen other elements that stand between a nuclear incident and a national catastrophe. It is his machinery that prevents a meltdown when everything else has failed.

For a moment, Dunn studies the graph of the accident that Kelly has pinned to the wall. Then he turns to them. "I'm concerned about these people turning off the high-pressure injection," he says. He points out that the emergency pumps were cut off by the operators about four minutes into the accident—at least fifteen minutes before they understood what was happening. "I can give you some scenarios where the situation would lead to possible fuel damage."

Kelly is stunned.

At Davis-Besse, they were operating at only 9 percent power. What if they had been running at full tilt?

3

Knoxville, October 1977

In a sense, Carl Michelson is a science-fiction writer. You've probably never read any of his stuff, but if you'd like to, you'll need a degree in reactor engineering to get through it. With a fertile imagination, fueled by meticulous research and vast practical knowledge, he looks into the future and writes scenarios for disaster. Fortunately for all of us, his best work has never been published.

In the nuclear industry his name is well known, and in the opinion of many in high places, he's a monumental pain in the ass. His reputation as a technological detective springs from an incident at Sandia Laboratories in 1975. He was in Albuquerque for a conference on nuclear safety, and at one of the sessions, he raised the question of sabotage. "What have we done to prevent somebody from damaging one of these plants intentionally?"

Everybody looked at him as if he were crazy. "Who would sabotage a plant? That's absurd. At the Tennessee nuclear-fuel facility we still have the honor system. People come and go as they please."

What about a madman?

"Anything some nut might do could be easily canceled by an operator."

Everyone present assumed that the subject had been dealt

with, but they soon discovered that Michelson is not your ordinary team player. At first glance, in his gray suit and wire-rimmed glasses, he seems like anything but a wave maker. But look closer; notice the set of the jaw. "I spent the next three weeks looking over documents that would be in the local public-document room and wrote a hundred-page report," he says.

When they read it at Sandia, there was shocked silence in the room. "There are at least eight different ways to do it," he told them. "That was all I had time for." Using documents accessible to anybody off the street, he presented detailed scenarios showing how a high-school graduate, without the use of tools or explosives, could cause a meltdown at almost any nuclear-power plant. "In fact you can train a janitor to do it. And there are things he can do to make sure it's irreversible."

Immediately the report was classified and all but two copies were destroyed. Michelson did not get a copy, but then he really didn't want one. Because of this secret document, tens of millions of dollars were spent upgrading security at nuclear installations all over the country.

And this is the sticking point with a guy like Carl Michelson. Whenever he does his job, it usually costs somebody a lot of money. And if he does his job too well, it may cost a fortune. He's what is known in the engineering trade as a "systems analyst." He's supposed to look at the whole picture and make sure somebody hasn't forgotten something obvious like the steering wheel.

Young engineers who have worked for him say it's an education. "There aren't any messenger boys on Michelson's crew. You've got to take a position. It was about the first time I ever had to think for myself."

He pushes his fledglings away from the desk and out into the plant, where they can get the feel of the hardware and the smell of trouble. Down there amid the swelter of pipes and compressors, he shows them how to shed the blinders of specialization and to take a look at the big picture with the simple logic of a plumber or a carpenter. "Does this look right?" Often the answer is no. Usually it's something obvious like the steering wheel.

The business about the containment vents, for example, was a fundamental flaw that eluded the designers, the government regulators, the builders, and the owners, until Michelson happened to glance up one day and ask a question. "Why's that vent open?"

"We keep 'em open all the time when we're running. Keeps gas from building up in the containment."

Michelson nodded. But in that second he had already seen a flash of disaster. The containment is the massive structure that caps the reactor, and it is usually built to withstand a force of several hundred thousand tons. But if you're running with the vents open, you might as well have used all that concrete for sidewalks.

True, the vents are designed to close automatically in an accident. But that's on the drawing boards. Out here in real life, any kind of serious explosion would blow all kinds of crap through those vents—shrapnel, loose insulation, who knows. Odds are, something would stick. And one stuck vent is all you need.

He decided to bypass his own management on this one. To them, the very notion of a serious explosion in the containment was almost too remote to deal with. One supervisor told him, "You can't think of everything, Carl. You've got to have a little faith in the machine."

But Michelson's faith is Lutheran; he's got no room for mechanical gods. He made an end run around his superiors and informally laid it before the safety experts of the Nuclear Regulatory Commission. "One NRC guy never understood what I was telling him. I don't know why."

The next one understood, but that was about all. "I don't

know what to do with this. I wish you hadn't told me."

Doggedly, he worked his way up to the man who was responsible for analysis of the containment design. Finally he was able to sit down with the man and ask him point-blank, "Did you fellas realize that maybe the containment isn't sealed at the time of an accident? And in fact, if an accident occurs, all the radioactive material in the containment building could shoot out into the atmosphere?"

"The guy said, 'Gee, I didn't know that.' And then he realized what was going on. All the same, it took about two or three years before the new regulations came out."

In many ways, Carl Michelson is the embodiment of his own icon: the professional engineer. Professional in the sense of, say, Roebling of the Brooklyn Bridge, or Brunel, builder of the British railway system. In the age of Cheops, we might have found him with chain and staff at Karnak or Giza. At the siege of Richmond, he would have been at the battle line, either building fortifications or planning their destruction. In the fall of 1977, he can be found in a spartan office in Knoxville working for the Tennessee Valley Authority.

TVA is one of the world's largest electric utilities. Builder, owner, and operator of some thirty hydroelectric dams and three operating nuclear plants with another fourteen on the drawing boards, it is a federal agency established by Roosevelt in the thirties, and it is one of the few major power companies in this country that are not in private hands. This means that Carl Michelson, no matter how much he may infuriate his superiors, cannot be fired. He is a civil servant.

This morning, he has a blueprint unrolled across the table in his office, and from the set of his lips it's apparent he doesn't like what he sees. It's not some detail that's caught his eye. It's the overall layout of the plant. It's so obvious, in fact, that for a moment he can't quite believe it.

For the past several months, Michelson has been analyzing the design of the Babcock & Wilcox nuclear reactor. TVA has two of them under construction in Alabama. What he's looking at now is sometimes called the "top sheet" of the blueprints, an undetailed sketch that shows where everything is. What worries him is the position of the pressurizer.

The pressurizer is a steel tank about forty feet tall that regulates the pressure in the main cooling system. It is one of the highest points in the main loop, and while every other part of the cooling system is filled with water, the pressurizer is topped with a bubble of steam that acts like a shock absorber. At the bottom of the pressurizer are electric heaters that can make the bubble expand. At the top are water sprays that can make it contract. Thus, the operators fine tune reactor pressure.

What is knitting Michelson's brow is the pipe that connects the pressurizer to the rest of the system. It enters the bottom of the pressurizer—as it must—but because of the layout of the other hardware, the pipe takes a dip, creating a U-shaped loop almost like a sink drain trap. And Michelson is enough of a plumber to know it would have the same effect as a sink drain trap. It would be a vapor lock. That's okay in a sink drain. It keeps the sewer gas from backing up. But here it's not so good.

Say, for example, that one of the safety valves on top of the pressurizer sticks open. The reactor will lose pressure; if the core is hot enough, the water there will flash to steam. But this steam won't be able to escape through the safety valves. It would be trapped in the main loop because of that dip in the pipe going into the pressurizer.

Worse, there is a possibility that the operators could be tricked into making the wrong decision. Steam in the core would force a slug of water up into the pressurizer. The operators might see the level rising and think the system is

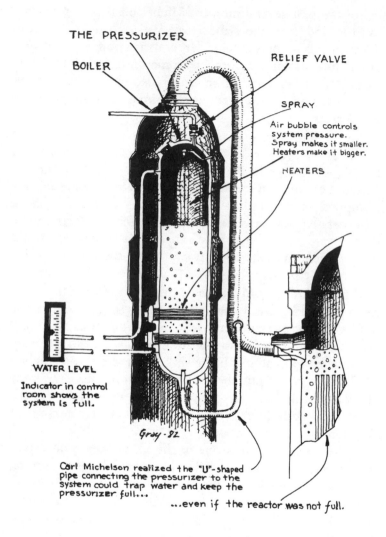

THE PRESSURIZER

BOILER

RELIEF VALVE

SPRAY

Air bubble controls
system pressure.
Spray makes it smaller.
Heaters make it bigger.

HEATERS

WATER LEVEL

Indicator in control
room shows the
system is full.

Gray - 82

Carl Michelson realized the "U"-shaped
pipe connecting the pressurizer to the
system could trap water and keep the
pressurizer full...

...even if the reactor was not full.

Figure 1.   The pressurizer. In normal operation, it acts as a
shock absorber for the reactor coolant system.

filling up when in fact it's emptying.

Michelson rocks back in his chair.

Too obvious. So obvious that it's going to be difficult to get anybody to take it seriously. He'll have to be careful with this one.

He has no way of knowing that the accident he fears has already happened.

Less than a month after the incident at Davis-Besse, Joe Kelly is sitting at his desk in Lynchburg when he gets a call from Toledo that brings him right out of his chair. There has been another accident, says Fred Faist. And while this event was not as serious as the September blowdown, the operators again cut off the high-pressure emergency pumps. In fact, this time they disabled the pumps before they even had a chance to start. It was almost as if they had an instinct for making the wrong move.

Bert Dunn's office is close by, and Kelly wastes no time getting there. It was Dunn who spotted the problem in the first place, and when the senior engineer hears about the latest incident, he's as distressed as Kelly. As far as Dunn is concerned, it's a bad idea to cut off the high-pressure injection under any circumstances, but to cut it off before you know what's going on is dancing with disaster. "For God's sake, let's get clear instructions out to the operators," says Dunn. "They somehow are misunderstanding this."

Kelly decides to find out what the hell is going on down in the training department. John Lind, the training director, is in the control room of the simulator when Kelly catches up with him. He first met Lind at Crystal River when the two men worked together on the start-up of the Florida Power Corporation reactor. Though they don't know each other all that well, they always got along okay, so Kelly comes right to the point. He lays out the scenario of the events at Davis-Besse and bluntly asks Lind if the operators are getting the wrong training in emergency procedures. "What are we teaching these people about when to secure the high-pressure injection?"

Lind is dismayed. He's at a loss to explain how the Davis-Besse operators could have blundered into forbidden territory. "I'm convinced we're giving them the right information. They're trained to watch the pressure and temperature as well as the water level. We teach them to keep HPI running until everything is stable."

There are several other instructors in the room, and in a few minutes they're all drawn into the discussion. Everything they tell him seems to support what Lind is saying. Kelly is relieved. But while walking back to his office, he realizes it still doesn't fit. On the one hand he's got absolute assurance that the operators are warned to keep their hands off the emergency pumps. On the other hand he has these two goddamn event reports. Somehow, somebody ain't gettin' the message.

Kelly decides to see if he can create some kind of forum to get this thing out in the open and see if there is really a problem here. Sitting at his desk, he thinks it over carefully. He doesn't want to sound like an alarmist. But he wants everybody to understand this could be serious.

NOV. 1, 1977

FROM J. J. KELLY, PLANT INTEGRATION

Two recent events at the Toledo site have pointed out that perhaps we are not giving our customer enough guidance on the operation of the high-pressure injection . . .

He points out that there are certain accidents that require continuous operation of the emergency system. He suggests a couple of guidelines for the operators and asks for everybody's thoughts on the matter. One copy goes to his boss; another to his boss's boss; a copy to Bert Dunn, of course; a copy to the head of training; and a couple of copies to customer service, since they're the guys who actually talk to the operators out in the field.

And ten days later when Kelly gets a response, it makes him cringe. It's a handwritten note—and slightly condescending at that—from a guy named Frank Walters, over in customer service.

> ". . . in the opinion of this writer, the operators at Toledo responded in the correct manner . . ."

Slowly, he reads it a second time.

> "My assumption and the training assumes that reactor pressure and level will tend in the same direction in the event of a loss-of-coolant accident . . ."

Well, that's obviously horse shit. Exactly the opposite just happened at Davis-Besse. The level shot right off the scale while the pressure was dropping like a loose toolbox. But what bothers Kelly even more is that this confused note is the only answer he gets. From the seven people the memo was addressed to, he hears not a whisper.

Somewhat painfully, Kelly has to face the fact that he must be too far down the ladder to get any action on this thing. Bert Dunn is going to have to get directly involved. As head of the emergency core-cooling section, Dunn surely swings enough weight to get things moving.

There are basically two separate arms of the organization here in Lynchburg. Kelly and Dunn are in the engineering section, where the nuclear plants are designed. The problem seems to be with the people over in customer service. Dunn agrees. "The people in the customer-service organization and the training organization are used to running power plants almost all the time the way a plant operates on a daily basis. And it's kind of like driving a car. You have one stud that comes out on one of the wheels on your car: you tighten down the rest of the bolts and you go on about your merry way. You don't particularly care. It's a nice practical solution.

"But if you should happen to come across a particular chuckhole, on a particular corner, when that particular stud is taking all the stress, you could bend that wheel. And as a result of bending the wheel—because you're in a corner —roll your car over and kill yourself. They're not used to thinking in that fashion in customer service." But Dunn and his people have no choice. If you cut the corner too close. in a nuclear plant, you might take a city full of people over the edge with you.

What worries Dunn—and Kelly as well—is the possibility of fuel damage. In one sense the fuel at Davis-Besse is the same as the fuel any other power plant. Like coal, it boils water to make steam to run the turbine generator. But the power here is on a scale that is difficult to imagine. It is created by the direct disintegration of matter into energy— a process only marginally understood by the world's best scientists.

Certain kinds of uranium atoms, when they are close enough to each other, will generate a spontaneous, ongoing process of mutual annihilation. When a loose neutron, flying at just the right speed, smashes into the nucleus of a uranium atom, the uranium splits. This division (fission) releases a phenomenal amount of energy, along with a couple of other neutrons. These two neutrons can split a pair of other atoms, releasing four more neutrons, in turn releasing eight, sixteen, thirty-two, sixty-four, and so on. If this chain reaction happens instantly, and if you have enough of the right kind of atoms in the package, you have an A-bomb.

There is no chance of such an explosion at a nuclear-power plant, however, because there isn't enough of the right kind of uranium present in the fuel. Nonetheless the reaction must be controlled somehow or it will quickly get out of hand and melt everything it touches.

Deep within the massive containment building at Davis-Besse, shielded from humanity by fifteen feet of concrete, the nuclear fuel is suspended in an awesome pressure vessel forty feet tall with walls of eight-inch forged steel. Midway between top and bottom hang 36,000 twelve-foot-long zir-conium tubes. Inside these slender tubes, in finger-sized pellets, are a hundred tons of enriched uranium. Each tiny pellet generates the same heat as a ton of coal.

This tremendous heat is carried away by a virtual Niagara of water flowing through the fuel pins at the rate of 350,000 gallons a minute. The coolant water acts only as a messen-ger. It travels only as far as the boilers, where it trades off heat to the water on the other side of the pipes, then rushes back into the core for another load of heat. Since it never leaves the reactor building, the outside world is protected even if the radioactivity should somehow leak into the pri-mary coolant water.

In normal operation, the reactor is held in check by a system of control rods that can be driven in and out of the core. These rods contain chemicals such as boron that eat neutrons. By absorbing loose neutrons they can slow down the reaction or stop it altogether. In an accident, these rods —without fail—must slam into the core to stop the chain reaction. To the operators this is known as the SCRAM.

The design of the Babcock & Wilcox automatic SCRAM is foolproof—as indeed it must be. If there is the slightest twitch in power to the control circuits, the jaws of the drive mechanism spring open like a gut-shot alligator and the rods simply drop vertically into the core. Motors may fail, but you can always count on gravity. And just in case you can't, each rod is helped along with a blast of nitrogen.

For the engineers in Bert Dunn's section, the problem begins at the moment of SCRAM. The chain reaction is over, but there is still tremendous heat in the core caused by unstable atomic fragments that continue to fly apart from then on. When uranium splits, it can turn into a variety of elements, but as luck would have it, only a few of them are stable. They decay into other elements that are equally unstable, and at each step they release a little zinger of energy.

Once the fuel goes critical, this process will continue for all time. Even after the reactor is shut down completely, this residual heat must be carried away or the core will melt. No one knows for certain what would happen after that, but some of the scenarios are quite frightening. It is Bert Dunn's job to see that we never have to find out.

If something should interrupt the flow of cooling water —say, for example, an earthquake that breaks all the main pipes—the core must be cooled anyhow. The alternative is unacceptable. To deal with this remote possibility—and dozens of others equally remote—Dunn and his people have designed an ingenious collection of pumps, compressors, sprays, and tanks known as the emergency core-cooling system. Patiently, over the years, using computer simulations and elaborate tests, they have explored every conceivable kind of accident and planned a way around it.

So it's understandable that Dunn is gravely concerned when he finds that the operators are shutting down the emergency systems in a way that could turn his calculations into oatmeal. But before he can make a move, another serious incident occurs in Toledo.

4

Toledo, December 1977

Jim Creswell sifts the stack of strip charts for the third time. The plot for pressurizer level is missing. "Strange," he thinks.

Creswell has come to Toledo to look into the transient of November 29, the third unplanned event to hit Davis-Besse in sixty days. This latest incident was caused by the simplest kind of mistake. A computer patch panel was miswired.

These panels are plug-in trays that function like a telephone switchboard: rows of holes connected with wires route information to various parts of the circuit. But one of the wires was in the wrong hole. When this tray was patched into the computer, the wiring error created a false demand for power. Immediately sensing something screwy in the works, the automatic safety systems took over, SCRAMed the reactor, shut down the turbine generator, and began a programmed sequence of bringing the plant to a halt. Everything functioned exactly as designed for about fifteen seconds. Then somebody decided to help it along.

As the turbine began winding down, one of the operators, an inexperienced guy who got jumpy, instinctively reached for the circuit breakers on the generator. His thought was to protect the electrical circuits from wild fluctuations, but on this plant, the switchover from selling electricity to buying electricity is supposed to be automatic. (The emergency systems must be kept running at all costs; when the plant stops making its own power, it must immediately start receiving power from somewhere else.) By interrupting the automatic sequence, the operator unwittingly cut off the outside world.

As he threw the switch, the whole plant was plunged into darkness. The only lights in the control room were the flashing annunciators on the control panel.

But this is exactly the kind of thing that people like Bert Dunn think about and plan for. Once again, the plant reacted automatically to protect itself. Down in the Auxiliary Building—the huge pump house adjacent to the containment—are a pair of 900-horsepower diesel generators. To start them, there is a forty-foot string of storage batteries. The whole system is completely independent, and either one of these generators is capable of running all essential systems in the plant.

This turned out to be fortunate indeed, because one of the generators immediately ran out of control and cut itself off. But the other one held its ground; the lights came back on and the plant was quickly stabilized.

Jim Creswell has, however, not been sent here to investigate the cause of this event. Quite the opposite. He has been dispatched to the site to see if the accident can be used in place of a safety test. In effect, the people at Davis-Besse want to use the data from the accident to prove that certain parts of the reactor worked as designed; that these elements, therefore, need no further testing.

This is an unusual approach, but in a way it's understandable. The testing process for a new reactor is complex, tedious, and expensive. After the plant is licensed, it is eased up to maximum power in careful steps over a period of months. At each stage, various systems are tested to see if they can hack it. Will they stand the pressure or will they come apart at the seams? Will the valve open on schedule or will it disintegrate into shrapnel?

These are questions you would not like to have answered on an unscheduled basis in the middle of the night. So they are systematically spread through a start-up process expected to take at least six months. But at Davis-Besse the process has been dragging. The loss-of-off-site-power test, in fact, had not been completed on November 29 when the reactor lost off-site power. The actual event suddenly became the test. But since the emergency system worked, Toledo Edison can't see much point in testing it again.

Jim Creswell is an inspector for the Nuclear Regulatory Commission—the federal agency that licenses these reactors—and it is his job to see that the power-plant operators play by the rules. It was generally understood by Creswell and the rest of the government inspectors that the testing process at Toledo Edison was being stretched because of a number of kinks in the Davis-Besse installation that needed working out.

This is not altogether uncommon, and it was dealt with in an atmosphere of understanding. The federal regulations governing nuclear power are almost as complex as the power plants themselves. The rule book has grown to several volumes over the last thirty years, and if you attempted to follow it to the letter you would never get anything done at all.

So, at the local level there has evolved an attitude that could be called "watchful protection." The regulators are watchful. But they also have some tolerance for the problems of the plant operators and a deep appreciation for the fifteen-year gauntlet that any utility must run in order to get a nuclear plant on line.

And at Creswell's office—Region Three headquarters in Chicago—they have perhaps as much understanding of the problem as any group of regulators in the country. One of

the first commercial reactors in the United States was built
for Commonwealth Edison on the Illinois River, and the
Chicago utility is now running nearly half its generators on
steam from uranium fission. In a sense, the utilities grew up
with the regulators, and they taught each other everything
they know.

For Creswell, frankly, the whole process seems a little too
chummy. True, the plants are monstrously complex and
difficult to get underway. All the more reason to stick by the
book.

Among his own colleagues back in Chicago opinion is
divided as to Creswell's effectiveness as an inspector. But
here in Toledo there is almost unanimous agreement that he
is a nit-picking son of a bitch. Like a cop who can't see the
difference between a parking ticket and a murder rap, this
guy has no sense of scale. You never know when he's likely
to enforce some obscure regulation that doesn't make a
damn bit of difference to anybody.

It's an attitude that ignores economic realities, and it got
Creswell into trouble almost as soon as he arrived in Chicago. One of his first assignments was an inspection of the
Prarie Island reactor in Minnesota; his brief journey there
shook the whole organization right down to the filing cabinets.

Creswell found that the training program for the so-called
unlicensed operators was, at best, disorganized. The unlicensed operators deal with the physical work in the plant—
whatever can't be handled from the control room—and
while they don't need a degree in physics, it's important that
they know a block valve when they see it.

Creswell felt that the situation was completely unacceptable, and he cited the company for violation of the Nuclear
Regulatory Code—the first such citation, incidentally, ever
issued in Region Three. This produced the equivalent of a
meltdown back in Chicago; Creswell was called in and

reeducated by the senior inspectors. "A citation is a big deal," he was told. "Some of them carry criminal penalties. You don't get cooperation by pounding on people with a baseball bat.

"This is not a war. We all want safety. And the way to get safety is cooperation, not confrontation. You tell them what you want and if they don't change, then do something about it. But you've got to give them a chance to clean up their act before you start swinging a club." Also, these citations are a matter of public record. The press tend to exaggerate anything they don't understand.

So the citation was quietly withdrawn.

Creswell is a southern boy, and he wasn't all that keen on coming to Chicago in the first place. He grew up in the Tennessee hills near Oak Ridge and got his engineering degree right close to home, in Knoxville. But the nuclear industry in the South ran into hard times, and a friend at the Nuclear Regulatory Commission told him the work was interesting and the pay was very good.

Somehow he convinced his wife that Chicago winters wouldn't be all that bad. They were young enough, they had their lives in front of them, and it would certainly be good experience. But after the business at Prarie Island, Creswell was aware of opinions floating around the office that he was "too abrasive." That kind of talk can be fairly crippling to a career in government. Not that he could get fired. It is virtually impossible to fire a civil-service employee. But he could be dispatched to limbo easily enough.

So Jim Creswell tries his best to become a Southern Gentleman: visibly polite *in extremis*, never raising his voice at meetings, all disagreements amicable. And he smiles a lot; but he sometimes smiles in disbelief. That's the real problem. He's totally transparent; if he doesn't believe you, it's written all over his face.

When he got to Toledo two days ago, he asked about the

missing data. Now his inspection is over, and as politely as possible, Creswell lets them know he's not leaving without the goods.

"Where's the analysis on the pressurizer level?"

This man has a lot to learn about playing on the team.

Reluctantly, the staff locates the missing data. At first glance, Creswell catches his breath. It looks as if the water level in the pressurizer dropped right off the scale. There's no reading at all.

When he points this out to the supervisors at Davis-Besse, they are unimpressed. Of course the water level dropped. That always happens when you SCRAM the plant. It cools off and the water shrinks. The point is that everything worked. They lost off-site power and the plant went into a normal cool-down without a hitch.

Creswell goes for the rule book. He looks up "General Design Criteria 13." It seems clear from reading it that the operators shall have a method of knowing the water level in the system at all times. He looks at Terry Murray. "How could they know the level if it dropped off the scale?"

Murray is exasperated. It seems futile to explain to this guy that losing the level indication for four or five minutes may be inconvenient, but it has no effect whatsoever on safety. People like Creswell would stick to the rules if they flew out the window.

They agree on a compromise. Toledo Edison will supply the NRC with a calculated estimate of the water level. They're confident it will show that the pressurizer was never completely empty.

Word of this encounter precedes the young inspector back to Chicago, but it is only a precursor, a black bird soaring in advance of the squall. The real storm boils up on his return.

Creswell was wrong about the winters in Chicago. They are a hell of a lot worse than he imagined. But his office, fortunately, is about forty minutes west of the Loop. At least he doesn't have to face that knife-edge wind that rips in off the lake and cuts the breath right out of your lungs. God knows it's bad enough out here in Glen Ellyn. And they say the winter of '77–'78 is going to be one of the worst.

Creswell parks his car and heads across the lot to Number 4, one of a cluster of brick-and-glass office buildings set among the gaunt and leafless trees. It's going to be just as cool inside, he suspects.

First he reports to Tom Tambling, the senior inspector for Davis-Besse. He explains the business about the pressurizer-level chart and tells him that in his opinion the incident of November 29 will not qualify as a systems test. The requirements for the test are quite specific. A bunch of the parameters have to be handled very precisely. This wasn't a test. It was an accident. The pressures and flow rates were all over the lot.

Tambling says nothing. But in short order, Creswell finds himself in Bill Little's office. Little is section chief in Region Three Inspection and Enforcement, and he is outraged. Now that Creswell has filed a negative report, the whole thing will have to be bounced up to the licensing branch in Bethesda. This will accomplish three things: further delay, further expense, and a whole new bunch of people having to look through this pile of crap to come to the conclusion Creswell should have come to in the first place.

This latter point is the salient one. If there is one unwritten rule in bureaucratic management it is "Do not let shit flow uphill." Little peels Creswell's hide for the better part of an hour. But everybody in the office knows Bill Little wouldn't be this upset all by himself. Basically, Little is a nice guy. If the issue is generating this much heat, it must be with the full cognizance of the chief inspector, Gaston Fiorelli.

Now, Creswell knows he is on a very short string. No

two-hour lunches. He redoubles his effort to be precise, thoughtful, considerate. But he can't keep his hands off the damn numbers. Digging through the files one day—out of curiosity more than anything else—he comes across the transient of September 24.

Creswell remembers hearing about it from Terry Harpster, the inspector who went over to Toledo to check it out. He's somewhat more of a team player than Creswell —in fact, Terry's never issued a citation—but he found the event of September to be pretty impressive nonetheless. The main thing that stuck in his mind was that one of the steam generators boiled completely dry in no time flat.

But Creswell is looking for something else. He wants to compare the pressurizer-level plot from this event with the incident of November 29 to see if——

What's this?

Right in the middle of a loss-of-coolant accident, they cut back high-pressure injection. Why in the name of sweet Jesus would they do that? They hadn't even figured out where the leak was. Judging by the level trace on the pressurizer, it must have been quite a picnic. He spreads out the data for a better look.

They were lucky. What if they'd been running at full tilt?

John Streeter is Creswell's immediate boss, and if his posture is evidence, he is anything but spineless. His ramrod bearing and short hair suggest a military rigidity that is offset somewhat by his sports shirts. Like Creswell, he is a southern boy with religious convictions, but where Streeter's convictions are fundamentally Christian, Creswell's are now best described as those of technological anxiety.

Unlike Creswell, Streeter is an optimist. He whistles while he works. Creswell can hear him coming down the hall. Tom Tambling, chief inspector for Davis-Besse, is also here. This little gathering has been called so that they could have a full and frank discussion—no holds barred—about whatever it is that Creswell is up to. Is it possible, for example, that he is digging into stuff that has already been dealt with and is, furthermore, not really in his department?

Tambling, who wrote the original inspection report, is a slow and deliberate guy. But when he's angry, the veins stand out on his forehead. He tells Creswell that in his opinion the operators acted in a timely and correct manner as outlined in the procedures. The water level recovered. What's the goddamn problem?

"We can't have an operator cutting off the emergency systems right in the middle of an event," says Creswell. "The guys in the control room have got to have clear, unmistakable orders that if HPI comes on, you leave it on until you're sure you don't have a leak."

But nothing happened, says Tambling.

"What if they were at full power?" he asks. "And what if they had an old core in the reactor?" The fuel at Davis-Besse was new; when the plant shut down, there wasn't much decay heat. But what happens, say, a year later when the core is full of radioactive garbage? Even after the reactor shuts itself down, there will be half a million watts of heat generated by the leftovers. Everything will happen a lot more quickly.

Streeter isn't convinced. Neither is Tambling. This kind of speculation is way out of line for people in Inspection and Enforcement. This thing was reviewed by headquarters. If there really had been a problem, the guys in Bethesda would have picked it up.

"I called Bethesda," says Creswell. "They don't have any documentation on this event at all."

Streeter thinks it over for a while. "Why don't you write something up? Let's take a look at it."

Creswell falls to it with a vengeance. After several reviews, extensive discussions, and delays, a message goes out from Region Three to the control room at Davis-Besse:

> Prior to securing HPI, insure that a leak does not exist in the pressurizer such as a safety valve or an electromagnetic relief valve stuck open.

A neat, simple directive. But when you're overloaded with work it's difficult to think of everything; distant problems tend to move to the bottom of the list. So the possibility that this information might be essential to the operators of the seven other Babcock & Wilcox plants is overlooked.

Anyway, chances are this problem is unique to Toledo. For the fact is, even among the Babcock & Wilcox plants, each installation is significantly different. In the first place, Babcock & Wilcox supplies only part of the facility. The structures, the control room, the turbines and generators— all that comes from somewhere else. And it is put together by yet another contractor—the architect-engineer—who may make fundamental changes on his own. Sometimes you will find two plants side by side that are very different.

For example, in eastern Pennsylvania, Metropolitan Edison has a pair of brand-new B & W reactors just south of Harrisburg on Three Mile Island. Unit One is a solid plant with a pretty good record. But the operators who work both plants all agree that Unit Two is a dog.

5

Knoxville, December 1977

"You really win the safety battle in the trenches," says Carl Michelson. And after twenty-seven years with the Tennessee Valley Authority, he is a master of trench warfare. "You have to play games to get things done. I'm the first to admit that I play games. Not technically dishonest games. Simple strategies."

He's concerned about credibility. You can't be wrong in public too often or they'll stop paying attention to you. His strategy this time is to fly the idea in from left field. It's an approach that will keep his hands clean, at least until he has a better grip on the size of the problem he's dealing with.

As luck would have it, his former boss, Jesse Ebersole, is now a member of the Advisory Committee on Reactor Safeguards. The ACRS is a watchdog group of scientists, engineers, and professors who look over the shoulder of the Nuclear Regulatory Commission to make sure they follow their own rules. The ACRS reviews the license application for every new reactor, and they can report directly to Congress if they don't like what they see.

As it happens, they are at this moment considering the construction permit for a new Babcock & Wilcox plant at

Pebble Springs on the Columbia River near Portland. If Ebersole agrees with Michelson that there are problems with the plant's design, it will be simple for him to drop a few questions into the hearings. At that point somebody will have to deal with the issue, but nobody will know exactly where it came from.

More important, Ebersole is an old friend. He and Michelson have been in the trenches together long enough to have a great deal of respect for each other's abilities. If Ebersole doesn't think it's a serious issue, Michelson will probably drop it.

His opening salvo is an informal note—handwritten—with a list of concerns about the safety analysis of Babcock & Wilcox plants. For one thing, there is that "U" in the pipe that connects the pressurizer to the main coolant loop. This "loop seal" could trap steam in the core, and this could trick the operators into thinking the system is full of water when it isn't.

A few days later, Michelson gets a call from Jesse Ebersole. Jesse got the letter and he agrees. It's possible, he says, that all these points are unreviewed safety items. Michelson nods. "I won't object if you want to go to the NRC with it." Ebersole says there's an upcoming hearing on the Pebble Springs reactor. He'll bring it up then.

So far, so good. And a few weeks later Ebersole follows up on his promise. During a break at the permit hearing, he catches up with Sandy Israel, one of the front-line supervisors at the Nuclear Regulatory Commission. Israel, however, is a pretty busy man, and this is, after all, a handwritten note. So, he glances over it. Nothing new here.

But back in Knoxville, Carl Michelson, cautious and thorough as always, is still on the case. Encouraged by Ebersole's reaction, he has decided to take this thing all the way. And he knows better than to leave it hanging on a handwritten note. Over the Christmas holidays, Michelson rechecks his calculations, expands on his original concerns, and types a formal report.

Section 4.6 of his memorandum is now quite specific:

> A full pressurizer may convince the operator to trip the HPI pump and watch for a subsequent loss of level. Although this response appears desirable, a full pressurizer may not always be a good indication of high water level in the reactor coolant system . . .

> The loop-seal configuration of the pressurizer surge line allows the pressurizer to remain filled as the reactor coolant system water level drops . . .

Early in January 1978, he fires off a copy to Jesse Ebersole. But at this point, strangely, it is almost as if the gods have had enough; fate starts dealing from the bottom of the deck. Jesse Ebersole's wife is gravely ill, and her illness is making great demands on his time. And in the middle of this, Ebersole himself is hospitalized for an old surgical wound. So the report lies unheeded on his desk.

But Carl Michelson is indefatigable. He leaves nothing to chance. He has also dispatched a copy of the letter to Babcock & Wilcox engineering headquarters in Lynchburg. And in time, Michelson's report filters through the organization to the desk of the man who can actually do something about it: Bert Dunn, chief engineer of the emergency core-cooling section.

Dunn is already familiar with the problem of the false water level in the pressurizer, because he knows something that Michelson doesn't. He knows that—exactly as Michelson fears—the operators at Davis-Besse have already been tricked into shutting off the high-pressure-injection pumps. For this reason, Dunn is certain the issue has been dealt with. He dealt with it himself.

As soon as he heard that Joe Kelly had run aground trying to get the procedures rewritten, Dunn fired off a blunt warning to his superiors:

FEB. 9, 1978

FROM BERT DUNN, EMERGENCY CORE COOLING SECTION

This memo addresses a serious concern within ECCS Analysis about the potential for operator action to terminate high-pressure injection following the initial stage of a LOCA.

. . . concern here rose out of the recent incident at Toledo. During the accident the operator terminated high-pressure injection due to an apparent system recovery indicated by high level within the pressurizer . . .

I believe it is fortunate that Toledo was at an extremely low power and extremely low burnup. Had this event occurred in a reactor at full power with other than insignificant burnup, it is quite possible, perhaps probable, that core uncovery and possible fuel damage would have resulted.

After a struggle with the people in customer service, he finally got them to agree on a carefully worded warning about cutting off the high-pressure-injection pumps. What he doesn't know is that the new instructions were never sent. The man whose job it is to forward the instructions is Frank Walters, the same man who arrested Joe Kelly's efforts.

Frank Walters has some concerns of his own. What if the emergency systems go off—as they often do—in response to a false alarm? Since there is no leak, the high-pressure pumps would quickly fill the system and start blowing water out the safety valves. That would create one hell of a mess.

It's important to understand the concept of housekeeping in a nuclear-power plant. When you shut down a facility like Davis-Besse, it means Toledo Edison is losing $20,000 an hour. If the cleanup takes a few days, they are out several million dollars.

It is Frank Walters's job to keep the customer happy, and the customer won't be happy if you tell him he's got to wipe out his safety valves every time there's a minor bump in the system. So Walters bounces the thing back to the people in engineering with a letter suggesting they rethink the whole problem.

Unfortunately, this isn't the only issue these men are dealing with at the moment. The truth is, the engineering department is drowning in a sea of paper. Because of growing public criticism of nuclear safety, the government regulations are in a state of flux; a river of technical paperwork is flowing out of the NRC, and it's rising by the hour. How do you sort the wheat from the chaff when a batch of inch-thick reports hits your desk every morning along with half a pound of technical memos?

The letter from Michelson is a good example. The damn thing runs twenty-six pages and raises questions that would require hours of computer time. But it's only a speck in the blizzard of paper they get from Knoxville alone. Since they started designing the plant in Alabama, they've received about 10,000 letters from the TVA, nearly 6,000 of them from the engineering group that Michelson works for; almost a third of these deal with questions of plant safety.

So when Frank Walter's memo finally reaches Bert Dunn's desk, it's not surprising that Dunn never sees it. For the next thirteen months, Bert Dunn and Joe Kelly assume that the plant operators have all been alerted. But they have not. And in the control rooms at Rancho Seco in California, at Oconee in South Carolina, at Crystal River in Florida, at Duke Power in North Carolina, and at Three Mile Island in Pennsylvania, they are still ignorant of the problem.

Then one day in the corridor near the vending machines, this ticking bomb comes within an inch of being defused.

Bruce Karrasch, who is Joe Kelly's boss, did, in fact, get a copy of Frank Walters's memo and he did manage to read it. But he never did anything about it. And Don Hallman, who is Frank Walters's boss, has been after Karrasch for months to get it settled one way or the other.

Karrasch is feeding coins into the coffee machine when Hallman spots him. They chat for a moment; then Hallman reminds him about Frank Walters's memo. Should these instructions go out or not?

"I have no problem with that," says Karrasch. Both men are on their way to meetings, so the conversation is cut short. But when Hallman gets back to his desk, he realizes that he doesn't understand exactly what Karrasch told him. "No problem" with what? Dunn's position? Or Walters's?

He reaches for the phone and calls Karrasch. No answer.

At about this same time, a strange event takes place in the offices at Lynchburg. Bert Dunn gets a call from J. T. Willse in customer service. There is going to be a big meeting here in a few days. All the utilities now operating Babcock & Wilcox reactors will be sending representatives. But nobody is quite sure what the meeting is about. It seems some NRC inspector—a guy out of Region Three in Chicago—has requested a formal investigation into the pressurizer water-level problem.

This is a first. No regional inspector has ever requested a review like this. Management is afraid it will set some kind of precedent, and they're determined to nip it in the bud. There is going to be a full-dress presentation by the company, complete with slides and flip charts. The day before the inspectors are due in from Chicago, the engineering staff rehearses the pitch. Trying to imagine what the NRC is after, they practice fielding questions from each other.

Along with the Babcock & Wilcox people, a dozen representatives from the various operating reactors are present. Tom Hilbish is here from Three Mile Island. And while he didn't have to come all the way across the country, as Ray Dietrich did, he's as puzzled and distressed as everybody else about the purpose of this affair. According to the letter from Region Three, they want information from each of the utilities about their experience with the water level in the pressurizer. Specifically, has anybody besides Toledo Edison ever had the level in the pressurizer drop off scale during an accident?

As a matter of fact, there were a couple of incidents in 1978 at Three Mile Island, where the level did drop off the scale. There was a transient at Unit Two in April and another in November; but the core always stayed covered so it was no big deal.

And Dietrich's plant at Rancho Seco had a heart thumper last March. Somebody was changing a light bulb in the control panel when it slipped out of his fingers and dropped inside the console. That shorted out the computer, blacked out the control room, and triggered the emergency systems; and the plant went through a hell of a jolt. It cooled down so fast that some of the engineers were afraid the steel might fracture. But here again, the core stayed covered, so nothing really happened.

The following morning bright and early, everybody is out at the plant on Old Forest Road. When the boys from the NRC finally show up, it turns out to be quite a surprise. The two investigators, Jim Foster and Joel Kohler, are astonished to discover a room full of people from all over the country—along with the cream of the technical staff from the engineering department—armed and ready for a slide show.

It's all a mistake, they confess. They have not come down here to investigate the loss of pressurizer level. Rather, they have come down here to find out whether Babcock & Wilcox dealt with the problem "in a timely manner." They had expected the utilities to answer their questions by mail, not by sending representatives to Lynchburg. Foster is embarrassed and apologetic. Before the meeting is over, he assures everybody he's quite satisfied they all followed the proper reporting procedures and there is nothing to worry about.

Present in the back of the room is another NRC inspector, Don Anderson, who happened to be in Lynchburg on other business and thought he should sit in on this meeting. He is quite stupefied by this whole performance. He decides to find out what the hell is going on. During the break he suggests they all have lunch together, have a nice, friendly chat. And when he gets them alone, he points out that Lynchburg is normally his turf and he'd like to know what this is all about.

Foster reassures him. This whole thing got blown out of proportion, he says. There's this inspector in Region Three who has been raising hell about an accident at Davis-Besse. The guy is a real troublemaker. "The only reason for this investigation is to shut him up."

Anderson is impressed. It sticks in his mind. Later that night he writes in his diary:

shut him up

But the man they are talking about is Jim Creswell; it seems everybody still has a lot to learn about this fellow.

A few days later in Chicago, a briefing is held at Region Three headquarters out on Roosevelt Road. Creswell is there, of course; the meeting is more or less in his honor. And Creswell's boss, John Streeter; altogether about eight people. When everybody gets settled, Foster explains that he and Kohler have been investigating Creswell's concerns now for several months; they've been to Toledo a couple of times; been to Lynchburg; looked over all the memos. And the bottom line is they can't find a problem. Everything seems shipshape.

Creswell slides off the desk and starts out the door. "I've got work to do," he says.

"Oh, no, you don't," says his boss. "You asked for this investigation and you're gonna hear the results." For an instant the room is frozen. Slowly, Creswell sits down.

He did indeed ask for this investigation. He had hoped to be a part of it, as a matter of fact. But word came down from Jim Keppler, the director of Region Three, that the investigation would be conducted by somebody more objective about Toledo Edison than Creswell. So they sent Jim Foster and Joel Kohler instead.

Both Foster and Kohler had trouble understanding exactly what Creswell was talking about. In Foster's case, this is not surprising, since he has no technical background. He studied psychology, not engineering. He was hired by the NRC as a professional investigator, and he formerly served the government as an air marshall in the antiskyjacking program. Kohler, on the other hand, is a licensed reactor inspector, but has had no experience with Babcock & Wilcox plants. His years of experience with the Nuclear Regulatory Commission have convinced him that team players get the job done.

Privately, Kohler has a theory about Creswell's problem with the people at Toledo Edison. "If I want something, I get it. I don't resort to a citation to get a piece of information, because that will be the last piece of information I will ever get from that licensee.

"Either he didn't request it in the right way or he got them very angry at his manner or his technique. They were not on his side. They were fighting him every step of the way. They were not going to give him anything they didn't have to. They were going to make his life very difficult. And that is a problem that a person who does not communicate effectively can have in an inspection assignment."

A communications problem. That's the crux of the issue. Creswell does not communicate effectively. That is because he is speaking an alien language.

The difference between Creswell and his superiors is subtle but decisive. Most of these men are competent engineers. But it's difficult for them to take the possibility of a reactor meltdown very seriously. A phrase that appears and reappears in industry statements provides a clue:

"In the unlikely event of a serious accident . . ."

If you're convinced something is unlikely, it's hard to get exercised about it. And if the odds are a million to one, as the NRC's own study indicates, then it's even harder to get upset about something that's less urgent than sheltering yourself from meteorites. Creswell, on the other hand, believes Davis-Besse could go up the stack tonight—taking all of them and the nuclear industry with it. The nightmare quality of this investigation gives him the feeling of a man running in glue.

Jim Keppler, the regional director, is in the men's room when Creswell enters. Keppler tries to be pleasant. "How'd the investigation go?"

Creswell looks at him. "You really want to know?"

Keppler controls himself. "I wouldn't ask if I didn't want to know."

"I still have some concerns."

Keppler says he thinks all Toledo Edison needs is a little more time to get their feet on the ground. "Davis-Besse is typical of most new power plants," he says. "They go through problems. There is a period of learning that seems to go on between the utility and the reactor, and this is particularly true with a new power plant. The problems at Davis-Besse during this first year probably are about the same as the other plants'."

Creswell knows that's wishful thinking. The other Babcock & Wilcox plants averaged about sixty events in the first year. At Davis-Besse, the figure is running three times that high.

Keppler suggests they get together and talk. Later.

This episode cements Creswell's conviction that Keppler and Fiorelli aren't going to take any action against Toledo Edison until the fuel starts melting. Unlike the others, he not only feels an accident is possible; he can smell it in the wind. In the history of technological disasters, the unsinkable ships, the indestructible bridges, the perfect flying machines were not undone by the obvious failure everyone had planned for, but by the unexpected lineup of details.

Like the lookout who sees some formless mass looming out of the fog, Creswell senses that the ship has been running too long on luck. Luck can change. That night he makes plans for the ultimate step. He is going to take this thing all the way to the top.

As it happens, there is a new face on the five-man commission that runs the NRC. A young New England lawyer— a man who once worked for Ralph Nader—has just been appointed by the president as a nod to the environmentalists. His name is Peter Bradford, and when Creswell read some of his testimony before Congress he had to admit the man made a certain amount of sense.

But this is the point of no return. He has the feeling of bridges burning beneath him. At the last moment he has some second thoughts. What if these guys are right? What if I'm just shooting off my mouth? He talks it over with his wife. She doesn't understand the machinery, but she understands him, and she can see it written all over his face. You can always find another job but you can't always find another conscience.

Besides, he has less to lose now. Since the Foster-Kohler investigation began, the office has been about as tense as a pig on a high wire. And he's picked up a couple of new credits. Before, he was just abrasive. Now he's abrasive and immature. Sensing that this moment might come, he has already laid the groundwork. Over the past couple of weeks he has put out feelers to Washington.

The next day at lunch hour Creswell hangs around the office, and when he's certain he's alone, he dials the number for headquarters. In a moment, Commissioner Bradford's technical assistant, Hugh Thompson, is on the line. Softly but forcefully, Creswell says he considers the situation urgent and he would like to see the commissioner right away.

Thompson tells him it will probably be several weeks—maybe longer—before Bradford can schedule a tour of the Chicago office.

That's okay, says Creswell. He'll come to Washington.

Thompson is impressed. Creswell suggests that one of the other commissioners, John Ahearne, might want to sit in on this. Ahearne has a reputation for asking tough technical questions. Thompson agrees to get things in motion as discreetly as possible.

Thus, the following Saturday, James Creswell, regional reactor inspector, is off to the nation's capital to chat with some gentlemen who serve at the pleasure of the president. He is not unlike a letter carrier on his way to lunch with the postmaster general.

Creswell tries to lose himself in the crowds at O'Hare but when he steps up to the ticket counter his heart stops. There is Joel Kohler and another guy from the office. They're taking the same plane. Everybody tries to be pleasant, but

Kohler can't help noticing the bulging briefcase Creswell carries.

Fortunately, they're not seated together, so Creswell has some time to think. These guys are going to the NRC staff headquarters, which is out in Bethesda just off Wisconsin Avenue. The commissioners have an office building on H Street three blocks from the White House.

As they wait to get off the plane at Washington National, Kohler suggests they all share a car. Fine, says Creswell. On the drive into suburban Washington, he tells them he's come here to check his personnel file. Since he was just passed over for promotion, this is a thin but plausible story. And as soon as he loses them in Bethesda, he ducks out a side entrance and catches a taxi for H Street.

At the other end of the line, things take a turn for the better. Hugh Thompson has apparently taken care of business. As Creswell is ushered into a carpeted eleventh-floor office, he finds Commissioner Peter Bradford, Commissioner John Ahearne, and both technical assistants waiting for him.

They are patient, attentive, and earnest. It is more than he could possibly have hoped for. Referring now and then to the foot-thick stack of documents he brought with him, Creswell does his level best to communicate effectively. In the bluntest terms, he tells these powerful men that the Davis-Besse reactor should be shut down immediately, all the operators retrained, all of the broken equipment repaired, and the whole machine retested before they are allowed to come back on the line.

For an hour and a half he is grilled by Thompson and Ahearne, and the deeper they get into it, the grimmer it gets. At one point, Thompson asks if Creswell is absolutely certain the pressurizer problem hasn't already been dealt with by somebody on the headquarters staff in Bethesda. Creswell says, "Why don't you check the files?"

Thompson excuses himself. A moment later he returns with a look of surprise. "There's nothing in the file on the September 24 transient."

Creswell explains that most of Toledo Edison's difficulties are connected to personnel problems. There has been a terrific turnover among the operating staff; they've lost a lot of good people. And the engineering group is working over-time on the construction of Unit Two; they are spread too thin to deal with the operators' complaints about Unit One. Management control over the test program has been sloppy at best, he says. There is evidence of jury rigging the systems to conduct tests. There is evidence that testing has been delayed to allow them to generate electricity. This means they're operating at significant power levels with untested systems.

Peter Bradford is a stranger to the halls of heavy technol-ogy. But he has a logical mind and he's not stupid. He has no way of evaluating any of this data. It's also possible that this gentleman from Chicago is a raving madman. But if he had to lay money on it, he would lay it on Creswell. It's all too obvious, too mundane, too ordinary not to be true.

Ahearne agrees. For good or ill, they've got to rip the sheet off this thing and find out what the hell is going on. But it's essential for Creswell to retain what little anonymity is left. Since Kohler spotted him on the plane, they'll be looking at him sideways back in Chicago as it is. If, sud-denly, two of the commissioners start making detailed inqui-ries into the pressurizer business at Davis-Besse, Creswell's previous rap sheet will look like a rave review.

Ahearne is thinking. He's already had some correspon-dence about the Toledo reactor. Perhaps this previous in-quiry could be carefully expanded to disguise the actual motive. Hugh Thompson agrees to draft a memo for Ahearne.

That night Jim Creswell takes a late flight out of Washington National bound for Chicago. He paid for the flight out of his own pocket. He flew to Washington on his day off. And since he was virtually caught in the act he will probably find himself in quicksand when he gets back to the office. But he is satisfied about one thing. There is at last a chance something will be done. He has no illusions, but the meeting with Bradford and Ahearne was a definite change of pace. In fact they both seemed pretty upset. Now their technical assistants are digging into it; it won't be long before they confirm it for themselves. Creswell orders a drink and settles into his seat. And thirty-five thousand feet below, the dark hills of eastern Pennsylvania lie waiting for the edge of spring.

Just last week Jeanette Mayes planted her garden; the peas should be coming up any day now. From her front porch she can see the broad waters of the great river she has known since she was a girl. Some of the best damn fishing in the country. And less than a thousand yards to the south, out on the island, she can see the massive cooling towers: hourglass skyscrapers rising into the night, their graceful shape outlined by the wink of aircraft warning lights.

At last everything is in place. An elaborate, improbable plot, intricate beyond the dreams of any fiction writer; the story turns to dust if any one of several dozen people simply read their mail. But the machine itself is intricate. And as every student of probability knows, each roll of the dice is a new roll.

It is Saturday, March 22, 1979.

6

Harrisburg, March 28, 1979

In the whispering predawn darkness of the Susquehanna Valley, the ancient river flows past the Civil War waterfront of the state capital, slowly south toward Chesapeake Bay.

Harrisburg is asleep this Wednesday morning. The only sound is the whining rig of a sleepless trucker gaining on the Pennsylvania Turnpike.

Here the river turns east past Steelton and Highspire, then rounds a bend and passes a three-mile-long sandspit that is about to take the center stage of history.

Rising above the shoreline trees are four towers of such gargantuan scale that they seem to have been set there by visitors from outer space. Sweeping upward for thirty stories with graceful precision, their hourglass shape is designed to squeeze the water out of rising steam.

Tonight there is a torrential rainstorm inside the two south towers, a cooling rainstorm that wrings the heat out of a million gallons of scalding water per minute.

The water is coming from a windowless green building with the dimensions of an aircraft carrier. Inside, three stories above the island, a massive Westinghouse steam turbine is turning its 900-megawatt generator at 97 percent rated capacity.

To the north is a bullet-shaped dome that would dominate

the landscape itself were it not surrounded by such oversized companions. This heavy bunker houses the boilers that run the plant, and its construction is an indication of the un-imaginable power it contains. Over twenty stories tall, its steel-shielded concrete walls reinforced with cables as thick as a man's arm, it is designed to withstand the direct impact of a jetliner.

At the moment, the mighty plant is in the hands of shift supervisor Bill Zewe. An alert young technician with navy experience in reactor operation, Zewe and his control-room operators are running "hot, straight, and normal," a condition that is minting money for the owners. Naturally, in any machine that covers several acres, some niggling details have to be dealt with. When the previous shift signed the log book over at eleven o'clock last night, they left him with a knotty little problem in one of the filter systems, and his boys still haven't been able to crack it.

Zewe runs a tight ship; any time a problem drags on for several hours, he likes to have a first-hand look. So, for the second time tonight, he is making his way down to level 281 of the turbine building.

In a structure of this kind it is fairly meaningless to desig-nate a "first floor" or a "second floor," since there is such a variety in elevation from one place to the next. So the operators tend to use the numbers from the construction blueprints; here each level is designated by its height above sea level. Thus, "level 281" is 281 feet above sea level. Since the island itself is 305 feet above sea level, this means Zewe is in the basement.

But what a basement. From the cavernous chamber he is descending into, Zewe can see 300 feet to the west and 180 feet to the south through a forest of I-beams and pumps and tanks. In the middle of this vast arena, completely indepen-dent of the rest of the structure, a massive concrete pedestal rises five stories through the center of the building to support the turbine and its billion-watt generator.

Everything and everyone here have but one ultimate purpose: to keep the turbine spinning. To this end, a quarter of a million pounds of high-pressure steam are arriving here from the reactor building each minute. Superheated to 600 degrees and almost explosively powerful, the steam thunders into the turbine housing, blasting through row on row of spinning blades with enough force to turn the block-long, 500-ton steel shaft at 1,800 revolutions a minute.

Then, its energy spent, the steam is drawn down into water-jacketed coolers the size of a house, where it condenses to liquid, ready for another trip to the reactor building. But before it can be sent back to the boilers, the water must be filtered. And this is the thing that's giving Bill Zewe and his crew a fit. Stretching before him along the north wall of level 281 is a row of squat, 2,500-gallon tanks—eight in all—filled with BB-size balls of resin that clean the water as it passes through.

On the shift before Zewe got here, the operators sealed off tank 7 so that they could flush out the dirty resin and replace it. But as usual, the resin stuck. The pressure of 5,000 tons of water an hour flowing through the system tends to mash the beads down; also, the flushing system is underdesigned.

The engineers first discovered this during plant start-up back in 1977. They tried to fix it then by running a high-pressure air line to each tank so that the operators could tweak the valve and give the beads a shot of air to fluff them up and get things moving.

But tweak as they might, they're getting nowhere. They've been at it now for nearly five hours and the goddamn tank's still jammed up tighter than a cookie jar full of raisins.

Bill Zewe may look like a youngster, but at thirty-three he's had plenty of experience with nuclear plants and this one in particular. Like most of the men on this shift, he came here from the nuclear navy. He was on the island while the plant was still under construction. Six years ago he got a job with Met Ed as an auxiliary operator, and he's been working his way up through the ranks ever since.

Zewe's responsibilities are awesome; they include not only Unit Two but Unit One as well. On the graveyard shift, there is only one supervisor for the whole installation. Fortunately, there's nothing happening in Unit One to-night; they've been down for refueling. Which is just as well; there's a mountain of paperwork waiting for him on the desk upstairs. Also, he's scheduled to meet one of the trainees at four A.M. for a walk-through of the turbine build-ing. So, he leaves a couple of his best men working on the jam-up in tank 7 and makes his way back to the control room.

If you take this trip very often you won't have to worry about your waistline. From here it's all uphill. There's an elevator, but it's at the far end of the building.

Eight flights of narrow steel stairs threaded upward be-tween roaring pumps, the thunder of steam lines, and high-voltage conduits, bring the supervisor out on level 331, the "operating level." Here all other noise melts away to the sound of the great turbine, a piercing crescendo echoing and re-echoing through the dim green vastness.

But as Zewe steps through the door in the north wall, he leaves the noise behind. And as he steps through the next door, he is in an atmosphere of surgical silence.

By any standard, the control rooms at Three Mile Island are immense. In Unit Two, a ninety-foot wall of gauges and lights arranged in a sweeping arc defines the brightly lit center of the action, and just inside this, a low console follows the same line. There are other panels and banks of instruments in the dimness beyond, and the room itself seems to go on forever.

Zewe gives the panels a cursory glance. Everything is too

far away for him to actually read the instruments, but one can get a rough idea of the shape of things by checking the alarm panels.

About forty or fifty of these alarms are always lit up—chronic malfunctions either in the alarm itself or the equipment it's monitoring. In some cases the alarm has not been set properly, and it comes on during normal operation. But you get so you can ignore these nuisance signals.

Looking around the control room, the crew on this shift seems as if they just stepped out of a Western movie. Zewe is the only one clean-shaven. The other three—Faust, Frederick, and Scheimann—have long sideburns and mustaches, and Scheimann in particular, with his steely eyes and drooping mustache, looks ready for a gunfight.

Zewe, on the other hand, looks like an airline pilot: clean-cut and square-jawed. And he thinks like an airline pilot: methodically. At the top of his list right now is that damned filter down in the turbine building. So he asks Scheimann, the foreman, to go down to 281 and see if he can give Miller a hand.

Zewe steps into his office at the rear of the control room and rummages around for his lunch pail. Hard to get used to lunch in the middle of the night. He finds a sandwich and sits down to sift through the stacks of paper. Through the window at the front of his office he can still keep an eye on the control room.

Out there, at the center console, is the man who is actually running the plant. And right now Ed Frederick is on the edge of his chair.

Frederick doesn't really like this control room very much. From his point of view, there are plenty of problems with the layout, and it doesn't take a genius to see there are too many instruments. A glance around the room shows over 600 alarm lights. From his swivel chair at the center desk,

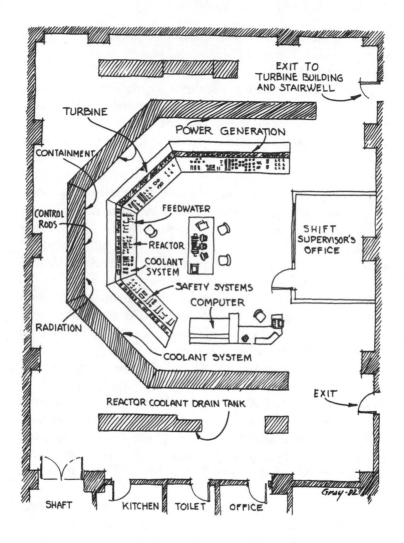

THE CONTROL ROOM

Figure 2. The control room at Three Mile Island Unit Two. To read the gauge for the level in the reactor coolant drain tank, the operators must walk around the end of the console to the second row of back panels.

Frederick can see 1,100 separate dials, gauges, and switch indicators.

Well, he can't actually see them, since a lot of the critical stuff is on the back panels over thirty feet away. And that's a problem in itself, since you have to walk about twenty paces around the center console to get close enough to read some of these dials. Sometimes it takes two operators to make an adjustment because the meter is on the back panel and the switch that controls it is on the center console. The only way you can deal with that is to put a man on the back panel to shout the numbers and another on the switch at the console.

The problem is that this place was designed for normal operation. In the middle of a dull afternoon, it's real handy to have a light come on and tell you about a problem on some pump that's three stories down and two blocks away. It means you can run the place from here with a lot fewer people.

But in a balls-out emergency, none of this is very helpful. If the system takes a jump, it usually shakes everything. In a typical shutdown, you may see twenty or thirty alarms go off at once. There doesn't seem to be any rhyme or reason to the way they laid out these goddamn lights. One of the most important alarms—reactor coolant pressure—is right next to the light that tells you the elevator is stuck in the turbine building.

One time—it was about a year ago, come to think of it—Frederick was on duty when the emergency systems went off and the plant went through a real heart thumper. Frederick was so impressed with the uselessness of the alarm system that he fired off a letter to the plant superintendent:

> The alarm system in the control room is so poorly designed that it contributes little in the analysis of a casualty. The other operators and myself have several suggestions on how to im-

prove our alarm system—perhaps we can discuss them some-time—preferably before the system as it is causes severe prob-lems.

It was fairly gutsy of Frederick to raise this issue, since a redesign of the control room would cost a fortune. But nothing ever came of the memo anyway, good or bad. It just disappeared somewhere up the ladder.

It still bugs him, but he's learned to live with it. Maybe he worries too much. Besides, Frederick has plenty of confidence in the guys around him. Between him and his bosses, Zewe and Scheimann, and his assistant, Craig Faust, they have over twenty-five years of navy training. If anything weird were about to happen, you couldn't ask for a better crew. They trained together on the Babcock & Wilcox simulator down at Lynchburg, and their test scores were near the top. Faust scored a little lower than the rest, but even he was above average.

It's a team you can count on in a pinch, and that's reassuring, because right now Frederick's got his hands full. For one thing, the relief valve on top of the pressurizer is leaking, which is a real pain in the ass. It means steam is blowing out of the main coolant loop all the time—not much, but enough so you have to make constant little adjustments to keep it in balance.

On the other hand, they've all had plenty of experience with it; the son of a bitch has been leaking for months. Naturally you can't shut down the whole place, at a million dollars a day, to fix one valve. So you put it on the list and you fix it during the next scheduled shutdown. That's what they're doing over in Unit One right now. In the meantime, it's a bitch for the operators. In a sense, Ed Frederick is not unlike the indefatigable Captain Alnut of *The African Queen*. The engine has a few idiosyncrasies but he knows where to kick it.

Deep within the bowels of the turbine building, however, is a ghost in the machine that Ed Frederick doesn't know about—a detail so infinitesimal that it's hard to imagine it could have any effect on this thundering giant. It is a small pipe, no bigger than a broom handle, connecting the plant's two compressed-air systems.

One of these systems, the "instrument air system," is used to open and close valves all over the plant. There is also a general-purpose air system for things like shooting bubbles into those filter tanks down in the turbine building. But during start-up, the engineers discovered that the general-purpose system wasn't big enough for the job, so they connected the two systems together.

This was probably not a good idea in the first place, but it was not inherently dangerous until a few hours ago. Then, somebody on the previous shift flushed out one of the filter tanks and forgot to close the air line when they had finished. Normally, it wouldn't make any difference, but the one-way check valve in the air line is leaking. For the past ten hours, water has been slowly inching its way up the pipe. It is now less than a foot from the air system that operates the valves on top of the filter tanks. If they lose air pressure, it's possible the valves could all close at once, instantly cutting off the supply of water to the boilers in the reactor building. The designers, however, were able to foresee this dangerous flaw, and they modified the controls so that the valves would not close but would fail "as is." Unfortunately the wiring for this modification was never connected.

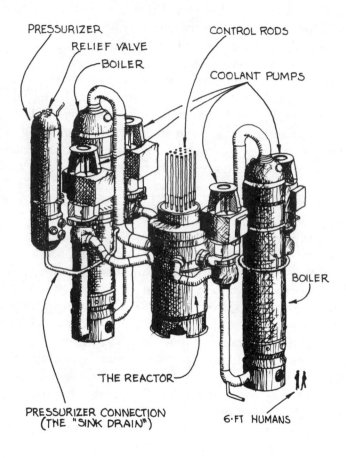

PRESSURIZER

RELIEF VALVE

BOILER

CONTROL RODS

COOLANT PUMPS

BOILER

THE REACTOR

PRESSURIZER CONNECTION
(THE "SINK DRAIN")

6·FT HUMANS

Figure 3.   The Babcock & Wilcox Model 177 nuclear reactor.

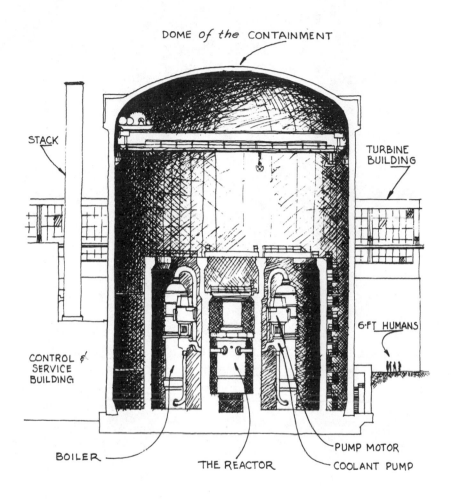

DOME *of the* CONTAINMENT

STACK

TURBINE BUILDING

6·FT HUMANS

CONTROL & SERVICE BUILDING

BOILER

THE REACTOR

PUMP MOTOR

COOLANT PUMP

Figure 4.   The reactor building at Three Mile Island.

7

And so she runs.

Black water sucked from the Susquehanna to carry away the wasted heat from ten million pounds of live steam per hour rushing through the tube sheets of the Westinghouse condensers, steam raining to water and drawn again into the cavernous reactor building, into the boilers glowing with the power of the universe itself, power of all power, the power of matter disintegrating into energy. A hundred and fifty tons of uranium, 25,000 grams of plutonium, one billion curies of iodine, 600 million curies of cesium, 680 million curies of xenon, 525 million curies of strontium hanging in slender rods, transforming itself in an alchemist's dream of nuclear violence, bathed in a thundering flow of water forced through the core by two-story pumps with power enough to reverse the Colorado River, carrying the heat of the cosmos upward through pipes conceived by giants for the giant task, heating the boilers that rise seven stories from the concrete vault—immense boilers, dwarfed by the building that contains them, monumental, rising dark and empty yet another seven stories to the dome, echoing with the scream and whine and shuddering roar of makeup pumps and recirc pumps, coolant pumps and sump pumps, oil pumps, steam seals, and valve motors, all in intricate

harmony working together to make steam. Steam superheated far beyond boiling, under the pressure of half a ton
to the square inch, liquid dynamite flowing to the turbine
building through wide-open throttle valves with a force of
seven million horsepower.

And so she runs.

An island of energy glistening in the night.

And here and there a human being, so small in this setting
that it's hard to imagine they are responsible for everything
here. Some sixty specialists in all, each with vast experience
and training, watching and listening for signs of change.

None is more experienced than Fred Scheimann. As he
descends eight flights into the pit of the turbine building, he
unconsciously checks each pump and valve as he passes by.
Any extraordinary noise, any leaking seal will go on a report
as soon as he gets back to the control room, and somebody
will be dispatched to deal with it.

When he hits the bottom step on level 281 he heads for
the filter tanks along the north wall. Don Miller is there, still
at the panel trying to jar the resin out of tank 7.

"Still constipated?"

"Won't budge," shouts Miller. "I'm using air, steam,
water. I don't think it's moved an inch."

At 3:58, Scheimann climbs up on the immense pipe running alongside the filter system so he can get a better look
at the sight glass on tank 7.

Hard to see the numbers in this light.

He takes off his glasses, rubs his eyes, and wipes the sweat
from his drooping mustache. Hot down here. Hot and
noisy. Hell of a lot——

Now what? Got awful quiet all of a sudden. One of the
pumps must have tripped.

"Hey, Miller——"

Then he hears it. A slug of water, propelled by steam,
moving like a freight train through the pipe. He jumps clear

just as the giant conduit leaps from its mounts, ripping out valve controls and fracturing one of the pumps in a spray of scalding water.

Eight flights up in the control room, Ed Frederick is rolling his swivel chair along the console to make another adjustment in the coolant system when something in Craig Faust's expression makes him spin around to see what the hell he's looking at.

"Something's happening," says Faust.

A high-pitched warble shatters the silence. Seconds later every alarm on panel 15 is flashing.

In the first instant, they tell us, you think of nothing. The mind is a blank. Then the brain engages in a preset sequence that was the point of all those hundreds of hours of training. Now Bill Zewe is out of his office, and it takes him only a second to size up the situation.

"Turbine trip."

"We lost the reactor."

Zewe's first thought is his responsibility to Unit One. They are getting power from Unit Two. He picks up the intercom to warn everybody at once.

"UNIT TWO. TURBINE TRIP. REACTOR TRIP."

Out in the turbine building, on the main deck, a couple of auxiliary operators are frozen in their tracks as the main safeties open, dumping a million pounds of unwanted steam into the night sky. Trainee Juanita Gingrich has never heard anything like this. It sounds as if the building is coming apart.

And out across the Susquehanna, in the fields beyond Goldsboro, a dairy farmer stops in the barnyard light to listen and wonder at the sounds of alarm over Three Mile Island.

The control room, however, is a scene of cool profession-

alism. Zewe, Frederick, and Faust have all been through this
before. Everybody knows exactly what to do. There's no
mystery about a turbine trip. But all these ass-kicking lights
and bells sure get the old adrenaline pumping. That siren
sounds like a goddamn gestapo raid. And every time somebody hits the button to cancel it, another alarm comes in.

"We've lost feed water."

"Verify emergency feed."

Any operator who's been through a loss-of-feed-water
trip on a Babcock & Wilcox reactor knows that the system
is in for a ride. The emergency feed water is several hundred
degrees cooler and when it hits the boilers it causes the water
to shrink in the reactor coolant loop. You have to work like
hell to maintain water level in the pressurizer, or the reactor
coolant will shrink right off the scale. There are rumors, in
fact, that in some Babcock & Wilcox plants, they run with
the block valves closed on the emergency pumps. It's against
the law. The NRC would shit if they found out. But the
truth is, the Babcock & Wilcox boilers are so touchy, you
sure as hell don't want emergency feed water dumped into
the system unless it's absolutely necessary.

Right now Frederick doesn't have any choice; the emergency feed pumps are already running. One of the first
things he has to do is get more water into the reactor loop.
But for some reason Faust is having trouble with the makeup
pump. Frederick lunges for the switch himself and gets it
going, so Faust moves away to cover the other panels. This
is the way they rehearsed it. One man on pressurizer to
watch level, one man on boilers and turbine. Zewe stands
back to watch for changes.

With two makeup pumps running, Frederick is still losing ground. The water level is down to 160 inches and
dropping. He hits the control valve to actuate high-pressure
injection. In effect, he's using the emergency cooling system
to help fill the reactor loop. And after a moment, the needle

on the gauge slows its descent and then holds.

The door bursts open and Fred Scheimann rushes in puffing and panting. It's taken him three minutes to reach the control room from level 281, and after a quick glance around the room he heads straight for the rack of emergency manuals. He flips to the procedures for turbine and reactor trip. Up to this point, Frederick and the others have been working by instinct. Scheimann wants to make sure nothing's been overlooked. Item by item, he shouts the emergency procedures as the operators verify what's been done.

There are pages of warnings and limits and instructions, and any one of them could lead to financial disaster if it's overlooked. The turbine, for example, is a sensitive piece of hardware. It must be slowed down carefully or you could get stuck with a seven-digit repair bill. And it must never come to a complete standstill while it's hot or the block-long steel shaft will sag and you'll be looking for other work. So when it finally spins down—a process that in itself takes an hour—you have to engage a turning gear that slowly rotates the shaft to keep it from warping. Unfortunately, the turning gear on Unit Two is broken, so in the middle of this confusion, Zewe has to dispatch an auxiliary operator out to the turbine building to turn the shaft with a jack.

By now, Ed Frederick is satisfied that the water level is on the rebound. As he moves to the feed-water panel, Scheimann steps in to cover the left wing of the console. That's the critical thing. Got to keep the water level up in the pressurizer. Scheimann shouts the numbers.

"We got 180 inches."

"A hundred ninety."

"Two hundred."

It's going up too fast. By the time Frederick works his way back over to the pressurizer panel, the level is up to 300 inches. They all know they must not take the system solid. Everything in the training, everything in the emergency

manual hammers that home. Don't take it solid. If the pressurizer fills completely, you'll blow water out the relief
valves, and they've never been tested for that. If you fracture
something in the reactor loop, you'll be in the middle of a
loss-of-coolant accident. Quickly Frederick and Scheimann
throttle back on the high-pressure-injection system, but the
level keeps rising. They shut down one of the makeup
pumps. Still it rises.

Three hundred forty inches.

Three hundred fifty inches.

For the first time, Frederick feels a twinge of something
more than excitement. This is screwy. Where the hell is this
water coming from?

"Three hundred eighty inches."

"Three hundred ninety inches."

"Okay, we're going solid!"

On the opposite wing of the console, Craig Faust is not
doing so well either. The boilers are going dry, which makes
no sense at all. The emergency feed-water system went on
in the first few seconds. By now the boilers should be filled,
but they're drying out.

Again, he verifies that the emergency pumps are running.
Again, he checks the valve lineup. EFW-8-A and B, open.
EFW-11-A and B, open. EFW——Wait a minute. For the
first time, Faust notices a yellow maintenance tag covering
up the indicator lights for valves 12-A and B. He shoves it
out of the way.

"The 'twelves' are closed!"

"What?"

"The block valves are closed!"

He slams the switches open. Over the noise monitors,
everyone can hear the machine-gunlike explosions inside
the boilers as the rush of cold water hits the superheated
tubes.

Already much more energy is trapped down here than the

operators realize. Anyone who's ever survived a close call is aware of the strange phenomenon known as time dilation. Everyone is susceptible, and Bill Zewe is no exception. He's under the impression they've been at this now for a couple of minutes, but in fact it's more like eight.

This is not good, because the boilers actually have two jobs. They boil water, but in the process, they also extract heat from the reactor coolant. Coolant flows inside the pipes; feed water flows outside. The feed water is under lower pressure, so it boils, and this sucks heat out of the coolant. When the boilers dry out, there is nothing to boil so the heat is trapped.

Heat on a scale that's hard to imagine. The nuclear fuel at Three Mile Island Two is dirty. The core has been operating for twelve months and it's accumulated a wealth of radioactive debris. Even though the reactor is shut down, the decay heat is stupendous. At the moment, the core is generating 30 million watts and getting hotter by the second. If this keeps up, the zirconium tubes holding the fuel will begin a chemical reaction with the cooling water. Nobody is quite sure what happens next.

It was for this very contingency that Bert Dunn and his group of engineers prepared so carefully. In the emergency core-cooling system, they created an interlocking network of emergency pumps and flood tanks designed to protect the core no matter what. After fifteen thousand engineering man-hours of testing and computer evaluation, Dunn hasn't any doubt that all the bases are covered. He has looked into the future and planned for the worst. What's more, the system is designed to function completely without human intervention. But in their determination to design the perfect machine, they have overlooked the possibility that humans might intervene anyway.

When the pressure in the reactor loop dropped through 1,600 pounds, the computer triggered the safety systems.

From Frederick's viewpoint, this is no help at all. The water level is already giving him a fit, and now all of a sudden the emergency pumps come on with another 1,000 gallons a minute. He is aware that it was the pressure drop that triggered the safety systems, but Frederick isn't thinking about the pressure right now. He's watching the water level climbing off the top of the scale.

In a moment of crisis, it's not uncommon for people to be hypnotized by a single instrument. Pilots lost in a storm sometimes become obsessed with the compass and fly right into the ground. Frederick isn't the only one who is transfixed by the rising water level in the pressurizer. Out of the corner of his eye, he can see half a dozen others milling around—a couple of engineers from Unit One, the supervisor from the other control room, several auxiliary operators waiting for orders—all looking over Frederick's shoulder at the water level slowly rising. One by one, he cuts off the emergency pumps until the flow of emergency cooling water is virtually stopped. Only then does the needle waver to a halt near the top of the scale. A wave of relief cuts through the tension.

But the pressure continues to drop and the temperature continues to rise, and for some reason, no one in the control room manages to put these two facts together. Zewe, for all his navy training and discipline, doesn't grasp it. Nor does Frederick. Or Faust. Or Scheimann. Yet they all know something is wrong.

"Call Jim Floyd and tell him to get his ass out here."

"Floyd isn't here. He's in training down in Lynchburg."

"Oh, yeah. Call George Kunder. And get hold of Joe Logan and tell him we've had a trip. Tell him we need some more eyes on this thing." In the pit of his stomach, Zewe knows he's missing something.

Right there in the drawer at the center console is a graph that holds the key—a steam table that shows the boiling point of water for any pressure. A glance at this graph would show the lines have crossed: the pressure is down to 1,200 pounds, the temperature climbing to 650, and deep within the reactor vessel, among the delicate tubes of fuel, it is boiling now.

How is it possible for four well-trained reactor operators to miss this most fundamental fact? Are they drunk? Certainly they're not stupid. Are they blind? Not necessarily. Try to imagine flying a jetliner full of passengers into La Guardia on a rainy night with your whole instrument panel flashing like a Christmas tree and the cockpit filled with bells and sirens. All things considered, everybody is doing his best, and it's almost good enough. Several times they come precious close to the answer. Even though he doesn't connect the pressure drop to the possibility of boiling in the core, Zewe is certainly aware that the pressure is lower than it should be. Could it be blowing out the power relief valve on top of the pressurizer? Maybe the valve stuck open. It's happened before.

Zewe asks one of the operators to check the temperature on the high side of the power relief valve. The operator punches a request into the computer and the numbers appear on the screen. The temperature is a blistering 283 degrees, a clear indication that the valve is blowing steam. But the operator looks at the number for the wrong valve. He gives Zewe a reading of 228 degrees. As far as Zewe is concerned, 228 degrees isn't all that bad. That valve, after all, has been leaking since January; the pipe is always hot on the upstream side. Another alarm comes in. He turns away. And in that instant the mighty plant is doomed.

For Bill Zewe, the immediate objective is to turn this thing around as quickly as possible. (1) Isolate the problem. (2) Restart the reactor. (3) Get the plant back on line. Every hour you waste costs the company about twenty grand. The pressure he feels is overwhelming. The rest of these people can concentrate on the event as it's unfolding, but he's got to figure out what caused it so he can get this mill back up to power and start making juice.

From up here in the control room, it looks as if the problem is somewhere down there on 281 in the turbine building. Those goddamn filters probably had something to do with it. He decides it's time for a first-hand look.

He shouts to Frederick, "How's the RC pressure?"

"Steady," says Frederick.

"Keep an eye on it. I'm going down to 281." A glance around the panel, and he is gone. Had he looked at the gauge himself, he would have seen that the pressure was "steady" at a mere 1,100 pounds.

If any one of dozen people had happened into the control room now, they would have seen in a flash what was happening. If Bert Dunn, say, or Joe Kelly from Babcock & Wilcox had been here, they would have told somebody to shut the block valve on top of the pressurizer, and this place would have remained an anonymous sandbar in the Susquehanna. But Kelly and Dunn are asleep in Lynchburg. Carl Michelson is asleep in Knoxville. Jim Creswell is fast asleep in Chicago. And their nightmare is in motion on Three Mile Island.

8

The power relief valve—Dresser Industries Model 31533VX-30—has a difficult task, and it is not equal to the occasion. Despite its sensitive position on the pressurizer at the highest point of the reactor-coolant loop, the valve is not considered a safety item since there is an emergency block valve just below it. So, its design was never reviewed by the government, and it contains a couple of flaws.

The sliding plug that opens and closes the valve has a tendency to stick, like a drawer that's too wide. And the coil spring that holds the plug in place naturally puts more pressure on one side than the other. So the valve is essentially failure prone. This same design has failed seven times before, once right here on the island. Tonight is number eight.

It opened, as designed, three seconds into the accident when the pressure shot up to 2,200 pounds. But when the reactor SCRAMed and the pressure dropped, the valve failed to close. So the reactor-coolant loop has a leak, the same as if a pipe had cracked, and what's more, it's a leak in an odd place. It causes the reactor to behave in a way the operators have never seen before, never read about, never even heard of. In every other trip, here or on the simulator in Lynchburg, the pressure and water level have moved in the same direction, rising or falling together. Now here is the water level rising and the pressure falling.

Their confusion is increased by an instrument that is lying to them, or at least telling only a half-truth. The red light on panel 3 shows that the power relief valve is closed, but it really means that the *switch* is closed. The valve itself is quite open, blowing 1,000 pounds of steam a minute into the drainage tank in the bottom of the containment building.

For fifteen minutes, a shrill river of steam has been blasting into this tank, and now it's full. As designed, the rupture disk blows out and the reactor coolant spills onto the floor. As this stream of liquid flows into a pit in the bottom of the building, a pair of pumps sense the rising water and kick on automatically. Normally, these pumps are lined up to an immense holding tank inside the containment building. But for some reason, the valves have been switched and the coolant is being pumped to a tank in the basement of the auxiliary building. This tank is already full and its rupture disk is already broken. Through this labyrinth of tanks and pumps, the steam blasting from the reactor now has a clear path out of the containment building.

Ed Frederick might have noticed that the drainage tank was filling if he had looked at the gauge on panel 29a, but to do that he would have to leave his post at the center console and take a walk around the end of the control room to the back panels. When Frederick finally sends somebody around in back to check it, the tank has already blown out and the level is back to normal.

Meanwhile, Bill Zewe has reached the bottom of the turbine building, and the situation is worse than he expected. The water hammer that shuddered through the system in the opening seconds of the accident has fractured a pump housing and there is water all over the place. And when the pipes jumped, they ripped out some valve controls; now the main condenser is flooded.

But Zewe is not a man to panic in heavy weather. The

navy taught him how to deal with damage control. After a quick look around, he rounds up a couple of operators and an instrument technician and puts them to work on the condenser valve. A couple of experienced hands are already at work trying to block off the fractured pump. As soon as he's satisfied that they've got it under control, he heads for the filter system along the north wall to see if he can figure out what the hell went wrong. He locates Don Miller, the operator who has been down here from the beginning, and Miller tells him that it all started when the valves went shut on the filter system. Zewe doesn't give a damn how it happened; he just wants to get the water moving again. There is a bypass around the filters, but he finds that the valve in that line is jammed shut.

He gets together a party of men and leads them up onto the platform surrounding the valve to see if they can man-handle it open.

"The goddamn wheel's missing."

"We can't find the hand wheel."

"Jesus H. Christ."

"Hey. There it is. Down there behind the duct."

"Get it up here!"

The hand wheel is carried up to the platform and fitted on the valve stem. Zewe and the others put their muscle to it but the valve won't budge. Zewe shouts down to the man on the telephone. "Tell Frederick to cut off the condensate pump. We gotta get the pressure off this thing."

With superhuman effort, they finally crack the valve open. Heaving with all their might, they manage to turn it until the motor can take over and the valve can be operated from the control room. One last check around and Zewe heads for the stairway, pounding eight flights up to the operating level. Once again, he is a victim of time dilation; he's been down here about twice as long as he realizes.

When he hits the control room, the place is starting to fill up. George Kunder, the technical supervisor, is here, and so are a dozen others. As people come in, Ed Frederick has been assigning them jobs; by now there is an operator on nearly every panel in the room. There are several telephone conversations going on at the communications desk, men are shouting numbers and procedures, and through all this, the alarms keep coming. It's getting difficult to hear yourself think. Zewe pulls Kunder into his office to go over the computer printouts and try to get some perspective on this thing.

In the bull pit of the control room, Ed Frederick is still trying to get control of the water level in the reactor loop, but nothing he does seems to make any difference. It's as if the goddamn thing has a mind of its own. Here's the gauge saying the system is full but the plant isn't behaving that way. Frederick is looking at the gauge with increasing suspicion. Maybe it's busted. Looking around the room, he spots an instrument technician. He says to the man, "We don't believe the pressurizer-level instrument. It says we're solid, but we're not reacting that way. It must be wrong."

The technician heads down below to take a reading closer to the source. A few minutes later, he's back. He asks Frederick a couple of questions. "How did it go up? Fast or slow? Did it peg out?" He goes to the computer and studies the printout. Then he turns to Frederick with the bad news. "That's your level," he says, pointing to the gauge on the panel.

All in all, this is a bad moment in the control room. Because they don't like what they're seeing, they are beginning to mistrust the instruments. It is the first sign of panic.

Suddenly, the chaos is split by a wailing siren.

"Look at this. We've got a fire alarm in the reactor building."

As soon as Frederick cancels the siren, it sounds again.

This time it's from the control building—the building they are standing in. Zewe walks around the console for a look at the back panel.

"Pressure's going up in the reactor building."

By now, it is not only the men in the control room who are gravely concerned. Some of the old hands out in the plant are beginning to see things they've never come up against. Terry Daugherty has worked for Metropolitan Edison since 1973. He was a machinist's mate on a nuclear submarine, and he came here immediately after he was discharged. He is the union shop steward for Local 563 of the International Brotherhood of Electrical Workers, and is one of the most experienced auxiliary operators in the plant.

Since the shit hit the fan about forty-five minutes ago, he has been constantly touring the turbine building to keep an eye on the crew. Some of the people on this shift have never been through a trip before, but they seem to be holding up okay. Now he wants to take a look over on the primary side of the plant. He enters the auxiliary building and heads down the hallway to the radiation-waste panel. As soon as he gets there he notices a couple of things that jar him. For one thing, both of the reactor-building sump pumps are running. He looks at the level indicator; it's off the scale.

And another thing. What's that noise? That buzzing over by the door? He goes over for a look and finds a hand frisker, an instrument used to check people for radioactive contamination. The alarm is sounding on this frisker. He picks it up. The needle is pegged on the high end of the scale. Daugherty switches to a higher scale. It jumps to 5,000 counts a minute.

Daugherty drops it on the desk. He steps back and rings up the control room.

"Ed, there's something wrong down here. I got both reactor-building sump pumps running and the level is

pegged high. It's off the scale at six feet. Another thing. I got
an RM 14 down here that's showing a lot of radiation all of
a sudden."

Up in the control room, Frederick calls Zewe over and
says, "That's Daugherty down on the rad-waste panel. He
says the reactor-building sump level is pegged high and both
sump pumps are running."

An auxiliary operator waiting for orders overhears this
shouted conversation and steps in. "Ed, I just worked the
primary side a couple nights ago, you know. I remember
high levels in the tanks. We've had a fair amount of water
in the miscellaneous tank all week."

Zewe says, "Knock off the sump pumps."

Frederick picks up the phone and rings Daugherty.
"Knock off the sump pumps. Pull the interlocks and make
sure they stay there." After a second, he adds, "And keep
an eye out for anything, Terry. It looks like we got something going on in the building."

It's been an hour now since this nightmare started. For a
while, it looked as if they had it under control. But for the
past thirty minutes they have been pummeled and whipsawed by a cascade of failures on a scale no one had ever
imagined. The level in the pressurizer is being held down
with great difficulty, the main condenser is out of business,
the reactor-building pressure and temperature are slowly
increasing, and every time Zewe looks at the printout on the
alarm computer, he wants to throw the damn thing down
the elevator. It is truly useless. The alarm printer takes four
seconds to type a line, and the alarms have been coming in
two or three times that fast. So the computer is running
thirty minutes behind.

Worst of all, the pressure continues to drop, and this has
ominous implications for the mighty pumps that cool the
core. These are enormous pumps. There are four of them
surrounding the reactor, each as tall as a two-story building,

and together they can move 350,000 gallons of water a minute. But they are not designed to pump steam, and as the pressure drops, there is more and more steam forming in the reactor coolant.

In the control room, the first indication of trouble comes from the vibration monitors. The pumps are starting to shake. Frederick gets out the operating curve, and he sees they are already near the limits. Zewe, Logan, and some of the engineers have been moving back and forth between the panel and the running discussion in the supervisor's office, and the consensus is that they should shut off one pump in each of the two reactor coolant loops and leave the other two pumps running as long as possible.

But very quickly, these pumps are in trouble as well. The coolant is now an uneven mixture of froth and water, and the thousand-horsepower pumps are alternately racing ahead and shuddering violently as they hit a slug of solid water. Even though Frederick and the others can't hear the deafening noise of the struggling pumps, they can see the battle well enough on the instruments in the control room. The vibration monitors, the flow indicators, and the ever decreasing pressure all show they are losing ground.

Zewe clenches his jaw. This is going to be a big deal. There's no easy way out from here. Somehow you've got to keep water circulating through the core. But you can't leave the main coolant pumps running and just let the bastards burn up. What if the vibration cracks a pipe in the reactor-coolant loop?

"We got high vibration. We got low flow. We got low amps."

"And we're down below the curve for net positive suction head."

It's a violation of procedures to operate the coolant pumps under these conditions. Zewe bites the bullet. "Cut 'em off," he says.

Outside, the sky is getting lighter, and as the first shadow of the river trees falls on the concrete fortress at Three Mile Island, nothing indicates that anything has changed. But the great plant is mortally wounded. Until this moment, Zewe, or Kunder, or Joe Logan, or Scheimann, or Frederick, or any one of two dozen other operators, engineers, and supervisors standing in the control room could have turned this thing around. If Bert Dunn had awakened out of a sound sleep and called up and said, "Shut the block valve on top of the pressurizer and open all the emergency pumps," they might have saved it. But no longer.

With all the pumps finally silent, Zewe and Frederick are hoping to cool the core by natural circulation. Since hot water rises and cold water falls, natural convection should keep the coolant slowly moving, and this by itself can be enough to carry away the heat. What they don't understand is that so much steam has already formed in the system that both coolant loops are completely blocked. Steam has collected in the high spots and no water can get past.

The top of the core is now uncovered. The decay-heat energy level is a staggering 26 million watts.

The fuel rods are made of Zircaloy, an alloy of zirconium. And while Zircaloy has a lot of useful characteristics as a fuel container—it doesn't steal neutrons from the chain reaction —it has some bad habits as well. Zirconium reacts chemically with water to form zirconium dioxide and hydrogen. At normal temperatures, this reaction doesn't amount to much. But as the temperature goes up, so does the reaction. And the reaction itself releases heat, driving everything even faster.

Now the situation has changed fundamentally. As long as the system contained only steam, there was always a possibility that something could be done to condense the bubbles by getting the temperature down. But now the chemical reaction in the core is creating clouds of free hydrogen.

They are rising to the top of the system, further blocking the flow, and the only way to get rid of it is to vent it out into the containment building. If this hydrogen comes in contact with a matching quantity of oxygen, it will explode.

But Zewe and the others don't know anything about this. In fact, from their standpoint, things are starting to look a little better. The temperature has stabilized. It seems to be holding steady at 570 degrees. Unfortunately, this is not the actual temperature but a technological fluke. The hot-leg temperature went off scale some time ago and so did the cold-leg temperature. Since the top and bottom readings are both out of sight, the needle sits in the middle of the gauge.

They are now less than sixty minutes from the beginning of a meltdown, a disaster with such staggering potential that only war survivors can grasp any sense of it. When the core reaches 5,000 degrees, it will begin to melt through the bottom of the plant and probably nothing can stop it. Not the eight-inch-thick forged steel pressure vessel, not the twenty-foot concrete foundation, probably nothing but the bedrock of the Susquehanna can slow the white-hot hundred-and-fifty-ton ball of cosmic fission. And when it first reaches ground water it will flash to steam and fracture the rock, blasting skyward around the base of the plant in great radioactive geysers that carry death in the wind.

A little before five A.M., Brian Mehler was rousted out of bed by a call from the control room at Three Mile Island. Mehler is the supervisor scheduled to take over on the next shift. One of the plant engineers is on the line. "We've had a turbine trip and a reactor trip in Unit Two."

Mehler gulps his coffee, throws on some clothes, and heads for the plant. As he approaches along the river, driving like a bat out of hell up Highway 421, he crests the ridge and can see from here that the plant is down. There is no steam over the south cooling towers. At the north gate, he flashes his ID and drives across the company bridge onto the island. Inside the security building, as he checks in and picks up his radiation badge, he gets his first clue that something big is in the wind. Nobody out here in the guardhouse knows what's happening, of course, but they can hear the frantic calls on the paging system. There is nothing more telling than fear of the unknown.

When he reaches the control room, Mehler is stunned by the contradiction of the instruments. Zewe, Frederick, and Scheimann are all huddled around the pressurizer panel, still fighting to keep the water level down. But something else catches Mehler's eye. The pressure in the coolant loop is down to 900 pounds, and the pressure in the building is going up. He has a hunch. He checks the temperature of the pressurizer relief valve. It seems too hot. Mehler leans over the console and says to Scheimann, "Shut the block valve on top of the pressurizer."

Immediately, the reactor pressure bottoms out and the building pressure stops rising. It has taken two hours and eighteen minutes, but at last they have found the leak. For the moment, Mehler has saved eastern Pennsylvania from desecration, but his discovery could not have come at a worse moment. What the fates give with one hand they often take with the other, and Mehler has unwittingly forged the final link in an incredible series of coincidences.

With the leak now closed off, it's essential to get emergency cooling water into the core to refill the system. But by now, the confusion in the control room is complete and no one at the console realizes what has happened. George Kunder, the plant engineer, might have caught the significance of the open block valve, but he is on the telephone in the supervisor's office at the back of the room.

In fact, he is in the middle of a conference call with three men who could easily crack the problem, and they are hot on the trail at this moment. Jack Herbein and Gary Miller are the two top technical people in charge of Three Mile Island, and they have been on the phone with Kunder for the last thirty minutes. On the line with them is Leland Rogers, the Babcock & Wilcox representative for the site. As Rogers listens to the numbers that Kunder is giving them, a bell rings in his head. Low pressure . . . high level . . .

"George? Is the block valve closed on top of the pressurizer?"

"I don't know. I'll check."

Kunder opens the office door and shouts to an operator standing outside, "See if they've got the block valve closed on the pressurizer."

A minute later Kunder relays the answer. "The block valve is closed, Leland."

If he had asked the question a moment earlier, the answer would have been "no" and Rogers would have unlocked the mystery. But it would take a fictional intellect like that of Holmes or Watson to realize that the next question is *How long has it been closed?* In the heat of the moment, it's not surprising that Rogers fails to ask it. And so, by a hair's breadth, the three top technical minds available have missed the essential fact. The block valve has been open for over two hours, and a quarter of a million pounds of coolant has blown out of the system.

Since the operators don't realize this, Mehler's discovery has made things worse in yet another way. Before, at least some of the heat could blast out into the containment building through the leaking valve. Now the leak is closed, all the heat is trapped, and the temperatures inside the core are rising dramatically.

Zewe and Frederick don't know this, however, because the core-temperature instruments only read up to 700 degrees. Above this, the computer prints out question marks, and it is printing them now.

```
TC141  ???
TC142  ???
TC143  ???
TC144  ???
```

By now it is difficult for anyone in the room to get above the situation, because so many apparently unrelated things are going on at once. And in the middle of all this, another unexpected problem has been slowly building. It's beginning to look as if the reactor is trying to restart.

It is boron that controls the reactor. Boron is a chemical that absorbs neutrons and stops the chain reaction. The control rods are filled with boron. A glance at the reactor panel confirms that all the control rods are still in place from the SCRAM. But to keep the reactor safely shut down, the operators also depend on a quantity of boron mixed into the coolant water. Some time ago, Frederick asked somebody to get a sample of the reactor coolant, and they came back with a boron count that was way low. He thinks—he hopes—this count is a mistake, and he asks for another one, and then another. Every sample shows much less boron in the system than there should be. And now there are signs of increasing radioactivity in the core.

Zewe is beginning to think there must be some hookup —some pipe, some valve, somewhere on those goddamn blueprints—that they've overlooked. And somehow, fresh water must be leaking into the reactor coolant and diluting the boron. That would account for the high water level as well.

Frederick calls Terry Daugherty on the paging system, and a moment later Daugherty calls in from the auxiliary building. Frederick tells him, "We're getting clean water

into the primary system. I want you to see if you can figure
out any way that's happening. Just look around and see if
you can come up with anything."

Daugherty has worked with Ed Frederick for a good long
while and he's never seen him lose his cool. But there is
something in his voice now that makes Daugherty uneasy.
What the hell does he mean, ". . . look around and see if you
can come up with anything"?

Daugherty clears the desk and unrolls the blueprints for
the makeup system. As he studies the diagrams looking for
any strange cross-connection that might have been overlooked, he hears one of the health-physics technicians coming down the hallway. The man carries a radiation detector
with a telescoping probe, and he is sweeping it from side to
side as he moves along the corridor. Daugherty doesn't like
the look of this.

"You finding anything?"

"We're reading five rads in the 305 makeup valve alley
and ten rads inside the door where you go into the makeup
pumps."

Another operator shows up, and Daugherty asks him to
come on a tour of the auxiliary building. Room by room,
one floor at a time, they look the place over and find nothing.
Then, as they are leaving, they come to a hole in the first
floor that looks down into the basement. Daugherty decides
to check it out. He leans over the railing and his grip tightens.

"Hey! We're getting water out of the floor drain!"

He runs to a phone and rings up the control room. "Ed,
the auxiliary-building sump is overflowing. We don't know
where it's coming from."

Daugherty gets off the phone and starts trying to get a fix
on the sump level. The man with him watches the rising
water in dismay. "Man oh man, you know, we're really
going to be crapped up around here."

Then, down the corridor, shouts of alarm. A radiation technician runs past.

"Get your stuff and get out!"

There are fifty people in the control room and everybody hears it at once. The radiation alarms are unmistakable. For a fleeting instant they are frozen in their tracks.

"High radiation in the aux building!"

"High radiation in the containment!"

Frederick turns to Zewe. "We should call Unit One and tell them to secure."

"Okay. This is a site emergency. We're declaring a site emergency."

Zewe picks up the intercom. He is as surprised at the words coming out of his mouth as are the dozens of others who hear him echoing over the island.

"THIS IS UNIT TWO. WE ARE DECLARING A SITE EMERGENCY. THIS IS NO DRILL."

A site emergency. Jesus Christ.

Zewe hits the switch. Reverberating over the island, klaxons and warning horns pierce every stairwell and pump room with a sound that chills the bones of many of these old navy men.

Action Stations.

Well, at least this son of a bitch won't sink.

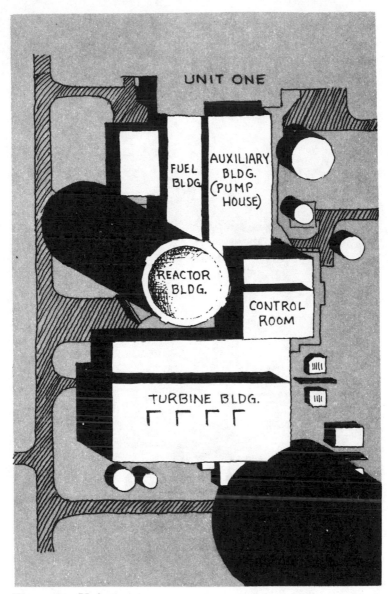

Figure 5.   Unit two.

Figure 6.   The island.

9

When Pete Velez was stationed in Scotland, he met a guy who was on the U.S.S. *Roosevelt* when it hit the bottom. It bounced off a peak in mid ocean. That's one of the worst things you can do in a submarine—hit something. If anything goes wrong under water, you're goin' down. Remember the *Thresher?* For a long time, they couldn't even find the pieces. It still makes Velez shudder. But you put those things out of your mind or you wouldn't be able to do your job.

The sun is up. Middletown is starting to show signs of life as Velez wheels down the main drag heading for the island. The Blue Room there on the corner is a saloon where the guys drink after work. From time to time you'll find Velez there at the bar, talking shop.

"Admiral Rickover, in my opinion, is one of the smartest men in the nuclear-power program. He knew how to run a program. He said this is the way I want it. This is the way you'll do it. And that's it."

Velez thinks that the civilian nuclear-power program is too loose. There are too many different designs, too many options. It's confusing for the guys who have to deal with it. Even plants that are supposed to be the same are completely different.

Heading south along 421, he sees the cooling towers rise above the river mist, and as he nears the island the plant covers the horizon like a cathedral. Normally, Velez is the kind of guy who can spot trouble a long way off; he grew up in Bedford-Stuyvesant in a Puerto Rican neighborhood surrounded by Italians and Irish. But this morning for some reason he fails to see it coming.

The north gate is closed and the cars are backed up, but that's no big deal. He figures they must have had another security intrusion. Most of them are just, like, a guy goes to close a door and doesn't close it all the way. You find the door open, you automatically have to call the security people until you verify that nobody's in there.

Velez pulls up to the guard and rolls down the window with a grin. "What's the matter, you lose somebody?"

"No."

"Well, what's goin' on?"

"Radiation emergency."

"*A drill?* At six o'clock in the morning?"

"It's no drill."

"Oops."

From the look on the man's face he can see it's no drill. Velez feels a twitch in his stomach. He gets out and heads for the phone in the guard's trailer. He dials the control room and gets through to the shift supervisor. "Hey, I'm out here at the north gate and I need permission to get on the island," he says. "Several of my technicians are here too."

"Come on in."

Driving over the long bridge to the island, he is thinking, "Thank God I practiced this. I know approximately what's going to happen and what I'm supposed to do. And I'm gonna do it. Now I better know what the hell I'm doin'."

Velez is a radiation-protection foreman. It's his job to see that nobody gets any unnecessary exposure—no more than 3 rem in three months except in an emergency. In order to save a piece of equipment that's vital to the plant, you may allow a man to pick up 25 rem. Up to 25 rem, you probably see very little blood changes, if any, and you probably wouldn't see any effect at all after a month or so. And it wouldn't do any major damage to the body. Or to save a man's life. How much is a man's life worth? They say 75 to 100 rem. But say the area has 120 rem. Do you go or not? It's a decision you have to make.

It's not that Velez is afraid of radiation. He respects it. Velez is a cautious kind of guy. In a gang fight in Brooklyn when he was with the Chaplains, he got an ice pick in the mouth, so he has sense enough not to take unnecessary chances. There are some guys on the radiation crew—guys like Janouski—who love to take chances. Janouski is a hell of a good technician, probably one of the best Velez has. But if there is a show to put on, you can count on Janouski.

Pounding up the stairs in the control building, adrenaline pumping, taking the steps two at a time, Velez is trying to think ahead. Portable radiation instruments are going to be a problem. They are short on instruments. The refueling over in Unit One used up a lot of the equipment. One of his first calls will be to some of the other plants to round up instruments.

Up to level 305, down the hall to the control room—what the hell is this? He opens the door. There's more noise inside than outside. People everywhere. Clusters of men all talking at once. Do these guys know what the hell they're doing? After a moment, Velez is able to see the pattern. Over by the back wall is the maintenance foreman with several of his boys, here are the engineers, there are the operators, over there behind the west panel is his boss, Dick Dubiel, and the rest of the radiation-protection people. Just like at football practice, the specialists are in bunches. Back there in the shift supervisor's office, all the heavyweights are getting together. And at the communications desk in the center of the room a couple of guys are on the horn letting the state and the county officials in on the fact that it's not just another day at the office.

In his hip pocket, Velez carries a little green notebook to keep a record of anything that strikes him as important. He happens to be starting a new book today. He takes it out, opens it to page one, and writes:

Aw shit

By now there are more navy people in the control room than on the bridge of the *Enterprise.* There is even a captain on deck. Joe Logan, the station superintendent, once commanded a squadron of nuclear submarines, so he is literally the SOPA—Senior Officer Present Aboard. But Logan has had the job only a few months, and a civilian plant like this is ten times as complicated as a navy reactor, so for all practical purposes, Logan is still in training. And he has enough sense not to get in the way. So there is little for him to do except stride the bridge and look calm.

But for Logan's boss, it's a different matter altogether. Gary Miller watched this place grow right out of the ground, and when he comes through the door at seven o'clock, nobody has the slightest doubt who is in charge.

Like Bill Zewe, Miller seems almost too young to deal with all this. But while he may have the unlined face of a college debater, he has been down to the sea in ships and has done business in great waters. Miller is the only commercial seaman here. He graduated from the Merchant Marine Academy at Kings Point, and after a series of berths as ship's engineer, he was offered a job in the navy yards at Newport News. In a matter of months, he worked his way up to being manager of construction on the nuclear carriers *Nimitz* and *Eisenhower.* He was barely thirty at the time, but his résumé was so impressive that Metropolitan Edison picked him to head up the acceptance test program for Three Mile Island. Gary Miller was the guy who actually signed for this place.

In the back of the room, Miller is immediately surrounded by the supervisors, who give him a quick sketch of what they know.

"We've got all four RC pumps shut down. We tried a restart twenty minutes ago and we got low amps and no flow. We got ES actuation. We got a radiation problem. We're getting activity in the hot-machine shop."

This is Bad News with a capital *B*. The hot-machine shop is where the chemists take samples of the reactor coolant. There is a small pipe that runs from the reactor, out of the containment building, and over to the sample room by the machine shop. If they are showing radiation over there, it's got to be coming from the coolant water. That means some of the fuel pins must have cracked.

Miller scans the radiation monitors. He wants to stay clear of Zewe and the people on the console, but even from the back here he can see that almost every alarm is lit.

Shit.

"Okay, everybody," he shouts. "As of now I am taking charge as emergency director. Let's clear the area. I want everybody who isn't working the panel to get back behind the lines."

This crowd is too big for him to deal with directly, so he creates an organization on the spot. "Mike Ross will be in charge of the plant," he shouts. "And Zewe, you report to him. Dick Dubiel will handle radiation protection. Dan Shovelin will head up emergency maintenance. Kunder is in charge of engineering. These people will all report to me."

All in all it is an impressive assemblage of talent. There are probably few other nuclear-power plants in the world that could have gathered this much technical ability on such short notice. Almost everyone here knows what is expected of him. They have rehearsed for this moment. The last drill was only a few months ago, and the government inspectors gave them a top rating. Miller even carries a set of index cards that outline his duties as emergency director. But he doesn't need index cards to know that his first responsibility is to the people of Pennsylvania. He tells George Kunder and Joe Logan to make sure all the government agencies are notified, along with company management and Babcock & Wilcox headquarters in Lynchburg.

When Kunder breaks out the yellow binders that contain the emergency plans, he finds them oddly reassuring. Here at last is a list of things to do. Mandatory phone calls to police and fire departments and federal agencies, orders to be given, repair parties to be assembled, radiation-monitoring teams to be dispatched. None of this has anything to do with the stricken reactor, however, and if any of these men could get a glimpse of what's happening in the core, they would be paralyzed. By now the water level in the main pressure vessel has dropped sixteen feet, and the upper half of the fuel is being cooled only by superheated steam from the violent boiling in the bottom of the vessel. In some places, the temperature has reached 4,000 degrees. This staggering heat has turned the surface of the Zircaloy fuel rods to a brittle powder and has generated astounding quantities of hydrogen in the process. Some of the fuel rods have cracked now and some have even melted, releasing radioactive xenon and iodine gas that is meant to stay trapped inside.

But Miller can't see the core; from here he can get only a hint of what's happening, and that's frightening enough. Radiation alarms are coming in constantly now and some of the numbers are starting to look scary. Inside the containment building, high in the apex of the dome, is a radiation detector that was designed for the worst event the engineers could imagine. The instrument itself sits inside a lead shield that cuts the radiation by a factor of a hundred. Right now this dome monitor reads 8 rem per hour. When you take the lead shield into account, that's 800 rem per hour, and it's rising as they watch.

Bill Zewe and the other operators at the console can see this, and so can everybody else in the room.

Eight hundred rem per hour is something they've heard about, read about, but nobody here ever expected to see it. Behind ashen faces, their minds race over the possibilities. But these men are navy disciplined. There will be no gut-wrenching screams of panic, no scramble for the exits. They have a job to do and they will do it.

"Eight hundred rem is the criterion for a general emergency."

Miller is shaken but clearly in control. "Okay," he says. "We're into a general emergency. I'm declaring a general emergency. Sound the alarms and notify the state."

On the front page of the big yellow manual Kunder is flipping through right now, a general emergency is defined as "an incident which has the potential for serious radiological consequences to the health and safety of the general public."

Kunder and the people on the phones have just barely finished notifying everybody about the site emergency that Zewe declared thirty minutes ago. Now they begin again at the top of the list.

"Dauphin County civil defense?"

"Yes."

"Yeah, this is Dick Dubiel from Three Mile Island nuclear station, okay? We have got a radiation emergency at Three Mile Island."

"Okay."

"Okay. Get in touch with Mr. Malloy——"

"He is here now. Would you like to speak with him?

"Yes. If you would, please."

Malloy gets on the line. "Yes?"

"Kevin? Dick Dubiel. Okay, we are in for real."

"Okay."

What else can you say?

Dick Dubiel and the radiation people have broken out the isopleths—wall-sized plastic maps of the area—and they are getting ready to plot the plume in case there is a release of radiation. As far as Miller can tell, nothing has escaped yet, but it's just a question of time. And the weather couldn't be more unforgiving. What you hope for at a time like this is a hurricane, something that will thin out the release and disperse it. Instead, they have mild winds out of the east— the worst imaginable situation. The town of Goldsboro, on the opposite shore, is less than 3,000 yards downwind, and there is a two-mile-an-hour breeze blowing right up Main Street.

Miller tells Dubiel to calculate a projected radiation dose for the people downwind, assuming the worst. There is a standard procedure for this, a formula that assumes a .5 percent-per-day leak in the containment. A moment later they're back with the numbers.

"We get a projection of 10 R per hour in the town of Goldsboro."

Miller knows this is a wild guess; he doesn't expect to find anything approaching this dose, but it's scary nonetheless. If the actual reading is even a fraction of this, they will have to evacuate immediately.

Miller studies the map and remembers thinking about this situation during the last drill. He realized then that the only answer would be a helicopter. The nearest bridge across the river is miles upstream, and now they're right in the middle of rush hour. He turns to Dubiel. "Tell the state police to get a chopper down here right away. We've got to put a radiation team across the river at Goldsboro."

The Coast Guard has to be notified in case they have to

stop traffic on the river. The airport has to be notified. They have to call Conrail and warn them that they may have to stop rail traffic past the plant. And the state police must be called to handle traffic on 441.

Here in the control room, the emergency plan seems to be running like clockwork. Out in the field, however, they are running into problems. Radiation protection has always been a stepsister of the industry—a necessary evil. Everybody gives it lip service, but it's pretty easy to forget about when everything is running smoothly.

So, in the security building, Pete Velez finds only four field-emergency monitoring kits, and when he unpacks them he finds that two are useless. One instrument is still in the repair shop, and somebody apparently dropped the other one into a trash compactor. So they jury rig as best they can. The helicopter still hasn't shown up. They decide to drive, but they can't locate a company truck. Finally one of the guys runs out to the lot and gets his own car, and they head off through morning traffic for the turnpike bridge.

The Honorable Richard L. Thornburgh has been governor of Pennsylvania only sixty-eight days, and he is already in the middle of a crisis. His new budget is under attack from every trench in the state house and he is fighting on all fronts. Thornburgh is a Republican—the kind of lean-and-hungry young Republican the party likes to point to with pride. Like all Republicans, he promised to cut the fat out of government, but to everyone's dismay, Thornburgh seems intent on slicing. For the past fifteen minutes, he's been on the phone defending the budget to a Philadelphia talk-show audience, and now he is on his way downstairs to glad-hand a bunch of freshman Democrats he's invited over for breakfast.

For many of the young legislators, this is a first look at the

inside of the governor's mansion, and as they eye it cove-
tously, Thornburgh moves among them with warmth and
ease. Then one of his aides appears. "Call for you, governor."

"I'll take it in the library."

On the line is Colonel Oran Henderson, the man whose
army career was scarred by the Mylai massacre. When Henderson retired, he found refuge in an obscure state post as
head of the Emergency Management Agency. Once again,
mysterious forces are inching him reluctantly to center
stage. "There's been an accident at the Three Mile Island
nuclear facility, sir. I'm taking the appropriate steps to notify everybody."

Standing alone in the elegant library, Thornburgh is immediately hit with the enormity of this snippet of information. There is no such thing as a trivial accident at a nuclear
facility. Evacuation is the first thing that crosses his mind.

"Find out everything you can and keep me informed. I
want to know the nature and extent and the dimensions of
this thing. Have you talked to the lieutenant governor?"

"Calling him now."

He hangs up and stares at the phone. After a moment, he
recovers his smile and goes back out to jolly up the Democrats.

"Captain Dave" doesn't like the sound of this. Dave Edwards is the traffic reporter for WKBO. Since the station
can't afford a helicopter, he cruises the freeways in his car
and scans the CB radio for signs of trouble. For the past
several minutes, he's been hearing an unusual amount of
action on the civil-defense frequencies. In Middletown, the
fire department and the police are mobilizing for some reason, but there isn't a clue about what's going on. Edwards
cuts off at the next exit and heads for Middletown to see for

himself. This is strange as hell. He radios the station and tells the news director, Mike Pintek, that something weird is going on.

Mike Pintek is an aggressive twenty-seven-year-old newsman with a certain amount of experience in the capital city. First he calls the Dauphin County civil defense to find out why they're mobilizing. Sorry, they tell him; don't know anything about it. Then he calls the police: same answer. The fire dispatcher: nothing. He's trying to figure out what to do next when Captain Dave radios in.

"There's nothing in Middletown, but out on the island, you know, there isn't any steam or anything coming from the cooling towers. The plant must be shut down for some reason."

Pintek dials the plant. He asks for somebody in public affairs. But the plant operator is so rattled by now that she witlessly connects Pintek directly to the control room.

"Unit Two."

"This is Mike Pintek, station WKBO. We heard——"

"I can't talk now. We got a problem."

Pintek is impressed. Everything he can hear in the background sounds pretty exciting. When the man hangs up, Pintek pulls down the directory for Reading, Pennsylvania, and dials the general offices of Metropolitan Edison. He asks for the communications director, Blain Fabian.

"He's in a meeting,"

"Get him out of the damn meeting."

Blain Fabian is at the starting gate of the worst day in his life. He has known about the accident for about an hour. Jack Herbein, the vice-president for generation, called him from Philadelphia at about 7:15 and together they drafted a two-sentence statement over the phone.

"The nuclear reactor at Three Mile Island Unit Two was shut down as prescribed when a malfunction related to a feed-water pump occurred about 4:00 A.M., Wednesday [March 28]. The entire unit was systematically shut down and will be out of service for about a week while equipment is checked and repairs made."

Fabian's operation is certainly not ready for this moment. His department is responsible for those little brochures they stuff into your electric bill. He's got a couple of writers that turn out an occasional press release, some graphic artists for the company house organ, and a few secretaries. The biggest thing they've ever handled is a rate increase. Now every line on the switchboard is flashing and they all want Blain Fabian.

"This is Mike Pintek from WKBO, Harrisburg. Can you tell us what's happening at Three Mile Island?"

Fabian has to be vague because he doesn't know anything. Since that first call from Jack Herbein, he hasn't even had a chance to go to the men's room, and nobody from the plant has bothered to call him with an update. He explains to Pintek that there is something they call a general emergency at the plant right now.

"What the hell is that?"

"A general emergency is a red-tape type of thing required by the NRC when certain conditions exist."

"What conditions?"

"There was a problem with a feed-water pump. The plant is shut down. We're working on it. There's no danger off-site. No danger to the general public."

As Pintek writes the story, he tries to tone it down so that people won't panic. At 8:25, sandwiched between a Top 40 hit and a lawn-care commercial, the citizens of Harrisburg learn that there is trouble, my friend, right here in River City.

10

Miles beyond the stately trees of Embassy Row, Wisconsin Avenue cleaves through a sea of office buildings, laundromats, and pizza parlors as it passes unceremoniously out of the District of Columbia into the Maryland suburb of Bethesda. Scattered over a six-block area near the line are half a dozen low-rise towers rented by the government for the staff of the Nuclear Regulatory Commission.

The NRC is the federal agency responsible for nuclear safety. It is a vast bureaucracy that sprang forth in full bloom —plucked out of the old Atomic Energy Commission— when Congress became concerned about a fundamental flaw in the structure of the original agency. The old AEC, which once had total responsibility for everything nuclear, whether civilian or military, was an outgrowth of the Manhattan Project, which developed the bomb during World War II.

Born in secrecy, with its own special language and a range of interests almost incomprehensible to the average citizen, the AEC ultimately developed a reputation for being the most arrogant bureaucracy in town. And in this town, that was saying something.

Critics in the House and Senate finally split the AEC apart when they realized there was a built-in conflict of interest in the law that set it up. This one agency, it seems, was supposed not only to regulate nuclear power but to promote it as well. As in any organization, the sales department ultimately carried the day. Staff members with legitimate concerns about safety were often shunted into the closet or shown the door to the alley.

In 1975, Congress sent the promoters and salesmen to the Energy Research and Development Agency and created the Nuclear Regulatory Commission to watch over nuclear power and make sure the plants were designed, built, and operated in a manner consistent with the public good. With its name thus changed, the regulatory wing of the old Atomic Energy Commission survived almost intact.

The five men chosen to run the NRC do not work in the grisly suburbs, but rather closer to the seat of power. Commissioners Hendrie, Bradford, Ahearne, Kennedy, and Gilinsky are presidential appointees; they are clustered on the top floor of an office building four blocks from the White House. Their dominion stretches from coast to coast, and in addition to the staff complex in Bethesda, they have five regional offices like the one outside Chicago where Jim Creswell works.

The regional office with direct responsibility for Three Mile Island is a hundred miles east of the stricken reactor, in the Philadelphia suburb of King of Prussia. When Bill Zewe declared a site emergency at 6:55, he was required by law to notify the Nuclear Regulatory Commission. He made the first call to Region One headquarters within fifteen minutes. The office wasn't open, so they left a message with the answering service.

When Gary Miller came on board and declared a general emergency at 7:24, a second call to the NRC was logged at 7:40. Again they got the answering service. But the operator has apparently sensed this isn't your ordinary business call. She looks up the number for Jim Devlin, the duty officer at

Region One, and calls him at home. Sorry, says his wife; Mr. Devlin has already left for work. The operator is getting edgy. She tries to signal Devlin on his beeper. Devlin gets the signal in his car as he's fighting through morning traffic, but since he's already on his way in, he decides it can wait.

Cruising down the turnpike, Jim Devlin is not the only one in blissful ignorance. The men in the control room are equally unable to grasp the scale of things. To the designers of this plant, it was inconceivable that the core could ever get hotter than 700 degrees. All their safety analysis, all their projections of accident sequences show this. So the computer is not programmed to read anything above 700 degrees, and the rattling teleprinter continues to hammer out row on row of question marks.

Miller knows he's got to get a better handle on this thing, and a good place to start would be to find out the temperature in the core. The instrumentation engineer, Ivan Porter, has been around, like Miller, since the place was under construction. If anybody knows a way to wire around that goddamn black box, it would be Porter.

Aside from the drone of people talking on every phone in the room, the shift supervisor's office is a relatively quiet little knot of tension. But the control room is a wall-to-wall madhouse. Miller has never seen so many people all talking at once. Again, he shouts above the din. "Let's hold it down! Everybody! Back of the line unless you're working the panel!"

The instrument guys have set up shop in a little office next to the control-room kitchen. When he locates Ivan Porter, Miller is calm, collected, but emphatic. "We've got to get a reading on the core. Is there any other way you can read those thermocouples?"

"I think we can."

Quickly, Porter rounds up a crew. One of the men says he knows which cabinet the wires pass through. He grabs a sheaf of blueprints, and with a couple of sidekicks, he heads out of the control room, down the tower steps, and into the relay room directly below. This is the chamber where the wires come together from all over the plant; it is a labyrinth. The lead technician unrolls the blueprints and directs the men to the computer-input cabinets. From the wiring diagrams they are able to identify the terminals that lead to the core thermocouples. They begin the delicate task of lifting the wires off and connecting their instruments.

Since the engineer, Porter, gave them no specific instructions about which thermocouples to read, they pick a few that are easy to reach and pull them off one at a time.

"Seven hundred degrees."

"Just over the line."

They check a couple more and get readings in the same neighborhood. Things are looking better. They hook up one more just to be on the safe side, and the needle jumps across the scale.

"Look at this."

"Twenty-two hundred?"

"That can't be right."

By the time Ivan Porter joins the men in the relay room, they have checked half a dozen terminals and the news is not good. A couple of readings are inexplicably low—around 200 degrees—but the rest are astonishing. On one set of terminals they get a reading of 3,700 degrees. The relay room is isolated from the cacophony upstairs; in the hollow silence, one of the technicians looks up at Porter.

"The core is uncovered."

"Bullshit!" says Porter. "I don't believe it. Are you sure you're reading that thing correctly? Have you got it hooked up right?"

Carefully, the man lifts the wires off another terminal and clips the instrument probes in place.

"Twenty-one hundred degrees."

"Something's the matter with your meter," says Porter. "Get a voltmeter and find out what the actual voltage is coming up from that thermocouple."

There is a scramble to find another instrument. Somebody manages to locate a digital voltmeter in the control room, and the foreman digs up a conversion chart so they can figure out the temperature once they know the voltage. Quickly they tap the circuits for the hottest readings. Their temples are pounding as they check the voltage against the chart.

"Forty-five millivolts . . . 2,100 degrees," he says.

"Same as before."

"Fifty-six millivolts . . . 2,600 degrees."

"That's impossible," says Porter. "That data's no good." He wheels out of the room and heads upstairs to find Gary Miller.

The men in the relay room look at each other. Right or wrong, these are numbers they've never even heard of. Maybe Porter can't believe the instruments because his mind simply refuses to deal with the information. He's got a lot invested in this place; he's been into nuclear power since he got out of college fifteen years ago. It might be easier for a priest to believe the Virgin Mother was a hooker than for Ivan Porter to believe that the fuel in his plant has reached the temperature of lava. But the men kneeling on the floor with the wires in their hands are losing their religious convictions.

"We've got a meltdown coming."

Originally it was Miller's plan to hold a meeting with his key people every thirty minutes or so in the shift supervisor's office, but the problems are so complicated that the first meeting melted into the second, and now it has become sort of a running session with Miller holding court as various people run in and out with the word.

When Ivan Porter comes in, Miller breaks off a conversation to get the report.

"It's all over the map. We've got a bunch of readings right around 600 to 800 degrees. We've got some that are way high. I mean 2,300 degrees."

"What do you think it means?"

"I don't think these numbers are reliable. We've got a couple of readings down around 200 degrees. You know that can't be right. And 2,300? I just don't see how that could be real. If you're asking for my personal evaluation, I think either the thermocouples or the lead wires have melted."

And that seems to be that. The readings are unbelievably high, so they are not believed. On close examination, Porter's explanation isn't all that reassuring. If the thermocouples in the core have melted, then what the hell is going on down there? But there's no time for introspection. Every phone in the joint is jumping off the hook, and they all want Gary Miller. Porter's argument seems to make sense; the readings are too erratic to be taken seriously. But as Miller turns away to pick up the phone, Ivan Porter feels the tiniest flutter in the pit of his stomach.

At a quarter to eight, the NRC regional office opens for business in King of Prussia. They turn on the switchboard, collect their messages, and in a few minutes collect their wits and open a direct phone line to the Unit Two control room. Although they are able to get only a sketch of the problem, it's not the kind of news you want to hear before coffee.

Logically, the word first reaches the headquarters staff by way of the public-relations department. The PR man in Region One calls the public-affairs director in Washington and tells him to brace himself because something's cooking at Three Mile Island. Within minutes, there are half a dozen phone conversations in progress between the regional office and headquarters.

"I'm getting you some information."

"Okay. I'll hold."

"Two hundred R per hour . . ."

"Yeah."

"Inside the containment——"

"Holy Jesus! Man, they got a release inside. Two hundred R per hour?"

The NRC, like Met Ed, has a lengthy list of emergency procedures, and one of their first steps is to activate the command post over in the East-West Towers. Known as the Incident Response Center, this complex of glass-walled conference rooms is set aside to function as a nerve center in the event of an accident. It is equipped with a gaggle of telephones, and the walls are lined with charts and screens for projecting things like plant blueprints and radiation maps. Key staff members are beginning to assemble from all over the organization, and a twenty-six-channel tape recorder is activated to monitor incoming calls.

What they need is hard data. Numbers. They are starved for numbers. And for the moment, the only tenuous line to the control room runs by way of the regional office in King of Prussia.

"Okay. Let me give you—presently, the initial contact we had about eight o'clock was reported as having up to 200 R per hour as measured by the containment dome monitor."

"Okay, slowly. Two hundred R——"

"R per hour."

"Inside containment."

"Inside containment . . ."

"Now, since that time we have two additional reports. One a little while ago of 600 R per hour——"

"Six hundred R per hour?"
" . . . the most recent one is 1,000 R per hour."
"One thousand?"

In the fertile valley of the Susquehanna are a dozen little hamlets with a dozen little newspapers whose reporters have to compete for stories working with no budget. A favorite low-cost beat is to phone the state police every once in a while and find out what's happening. A reporter from Waynesboro is making his morning calls when he hears about the problem at Three Mile Island. He tips the Associated Press bureau in Philadelphia and they shoot an inquiry to Harrisburg. The AP man in Harrisburg can't get through to Met Ed—the lines are all tied up—so he calls the state police and they confirm the story. But they have no idea what a "general emergency" actually is.

At 8:28 A.M., the bells ring in press rooms all over the country, and hundreds of teleprinters clatter in unison as AP moves the first 6 lines.

> HARRISBURG, PA.:—Officials at the Three Mile Island
> Nuclear Power Plant have declared a "general emergency,"
> a state police spokesman has said today.
> "Whatever it is, it is contained in the 'second' nuclear
> unit," said spokesman James Cox.
> "They said there is no radiation leak."

Not all of the radiation technicians who work for Pete Velez share his enthusiasm for the job. Tom Thompson, for example, thinks this place is the pits. He has little faith in his supervisors or his training. But then Thompson is not a navy man. He got here by way of a college degree in chemistry, and he doesn't particularly give a damn about nuclear power one way or the other. He studied psychology, in fact, and after being turned off by a short stint in his chosen profession, he decided to cash in on his science background. He applied for a job here as chem analyst in 1973. At least the money is good.

Thompson was due on shift at seven o'clock. He drove across the bridge right after Velez came through. As soon as he reported in, they sent him out with an off-site radiation party, but the instrument he was supposed to use was defective, so they sent him back to the radiation-control center in Unit One. Even in the drills, there was always a certain amount of confusion among the radiation people, but Thompson has never seen anything like this.

According to the emergency plan, the health-physics and chemistry people are supposed to set up a radiation-control center in whichever plant is not affected by the accident. So, Thompson's boss has set up the control point in the health-physics lab in Unit One. Like everybody else here, Thompson is pretty excited, but at least there's plenty to do. People are coming in from all over the plant crapped up with radiation. Most of it is gaseous, trapped in their clothes, and it dissipates pretty quickly. But it gives you plenty to think about. Thompson runs a frisker over one man and the instrument goes wild.

"Drop your clothes right here," he orders. "Get in there and shower."

Just then one of the foremen turns away from the phone and motions to Thompson. "We've got to get a reactor-coolant sample from Unit Two!" he shouts.

For an instant, Thompson considers quitting on the spot. He doesn't need a goddamn crystal ball to tell him that something major is happening over in Unit Two. There is no way to get a coolant sample this late in the game without releasing a humongous blast of radiation. "That sounds risky," he says.

"I said get it."

"You've got to be kidding," says Thompson. "Taking a sample down here—we're going to wipe out this control point." Thompson tries to explain that when they open the sample to analyze it, the lab will be filled with airborne radioactivity. But the man's in no mood for an argument.

"We've got to have that sample!"

Like most of us, Thompson would rather die than make an ass of himself. Muttering under his breath, he trudges off to the locker and tugs on some coveralls and a radiation wet suit. These people must be insane. He pulls a Scott Air Pack over his head and adjusts the breathing regulator. A few minutes later, he finds himself standing in a spot he never imagined in his worst nightmare.

Just down the hall from the health-physics lab is the hot-machine shop. Into this chamber come the sample lines from both reactors. Now Thompson is going to open the door to the hot-machine shop, walk in, and open the valve in a three-eighths-inch stainless-steel pipe that leads directly to the core coolant in Unit Two. Absolute fucking madness.

The air pack should be good for a protection factor of a thousand, he figures. Could it be worse than that? He tightens the face mask. One of the other health-physics technicians cracks the door and rushes in for a survey. He's out in a flash. "It's 200 R at the panel where you take the sample," he says.

Well, shit.

"This won't take long, boys."

Thompson hits the door running and heads straight for the panel. He connects the container, cracks the valve for an instant, grabs the sample, and runs out, all in less than ten seconds. Without missing a beat, he rips off his air pack and wet suit, then grabs for his pocket dosimeter.

Two hundred millirem.

Thompson resolves to get out of this chicken-shit outfit

at the next available opportunity. Immediately the area is rocked with radiation alarms.

Like a herd of gun-shy zebra, everyone in the room changes direction at once. And in the blink of an eye, the place is empty. Books, telephones, clothing, and coffee cups lie where they fell. What was a moment ago the emergency control center is now the deserted bridge of a mystery ship.

In Pennsylvania, Scranton is many things. In addition to being one of the oldest cities, it is also the name of one of the oldest families, as well as of the political dynasty that grew from it. The current heir is number-two man in the state house.

His father was himself governor and once a serious presidential contender, but if there was ever a potential president in the family it is thirty-one-year-old William Scranton III. Darkly handsome, tall, square-jawed, he is a candidate to make the admen weep. Fortunately for the people of eastern Pennsylvania, it turns out he also has a great deal of common sense—something in short supply this morning.

As lieutenant governor, Bill Scranton is the titular head of the State Emergency Management Agency. When it comes to disasters, Pennsylvania has plenty of experience. The mechanism for evacuation—from flood, fire, or mine explosion—always lies in wait like a coiled spring.

Although Scranton and Thornburgh have been in office for only a matter of weeks, the lieutenant governor has already had several meetings with the emergency-management people about the routine problems of springtime. So when he walked into his office shortly after eight, he assumed that the message from Colonel Henderson was about some stream that had jumped its banks.

He is stunned by what Henderson tells him—not only by the general emergency, but by how little anybody knows about it. How did it happen? Is it over? Are people in danger? Henderson can't answer.

Within minutes, Scranton has five phone lines tied up as he tries to reach anybody who might shed light on this thing. He calls the science advisor to the previous governor. He calls the State Bureau of Radiation Protection, the Department of Environmental Resources. He calls Met Ed. And finally he calls the control room at Three Mile Island.

Unlike most politicians, Bill Scranton is not ignorant of the technical issues surrounding nuclear power. When he was in his early twenties, he ran several of the family newspapers outside the city named for his great-great-grandfather, and his editorials were fairly critical of nuclear power. He didn't like the downside risk. Then, after the Arab oil boycott, he felt there was less room for the luxury of dissent. But he has read books like *We Almost Lost Detroit,* and he is aware of the tremendous potential for destruction.

As a former newsman, Scranton knows he's got to get out in front of this thing. Newspapers don't print blank space. He's got to tell them what he knows, sketchy though it may be. The gentlemen of the fourth estate are already in the state house, jammed into the press room for what they think will be his statement on the subject of energy. Jesus Christ, are they in for a surprise.

As the moment nears, word ripples through the crowd that the subject of the press conference has been changed; Scranton will discuss the accident at Three Mile Island.

"What accident at Three Mile Island?"

Suddenly, Scranton enters the room with a handful of advisors. The TV crews hit the lights, cameras roll, and he steps to the mike.

"The Metropolitan Edison Company has informed us that there has been an incident at Three Mile Island Unit Two. Everything is under control. There is and was no danger to public health and safety." Scranton tries to answer their questions as well as he can, but he doesn't know all that much, and besides, the terminology is Greek to everybody. What the hell is a "general emergency"? These guys are political reporters; they know less about that plant than the average tourist does. One reporter wants to know if any radiation escaped. Radiation is something everybody's heard of.

"There was a small release," says Scranton, but "no increase in normal radiation levels has been detected."

Then, unexpectedly, one of his aides steps to the mike. "They have detected a small amount of radioactive iodine," he says. The information came in too late to brief the lieutenant governor. It was only a trace, nothing harmful, but it was detected across the river in Goldsboro less than an hour ago.

A fleeting shadow, this unwelcome news narrows the eye. Scranton resolves not to get caught dangling again.

From all over the plant, the radiation alarms have been cascading into the control room. The auxiliary building has been evacuated, and now activity is starting to show up in the control building. The instrument at the dome of the containment is indicating 30,000 R per hour. A man might survive briefly in such an atmosphere, but he would be dead for all practical purposes in twenty-four seconds. This is a level of radiation that would transmute your wedding ring, and turn the gold in your teeth to lead. It is such a staggering number that everyone finds it impossible to believe. Maybe the instrument got saturated.

And then, Miller hears the alarm they have all been dreading: there is radiation in the control room. A quick sample of the atmosphere they are breathing indicates the presence of airborne radioactive particles.

Radiation, as Pete Velez is fond of saying, is like cow shit.

"If you smell it, that's radiation. But if you step in it, that's contamination. You can't get away from it. You carry it with you." If there is one thing the human body doesn't need, it is radioactive particles that stick in the lungs. At 10:17, Miller orders his men to break out the emergency breathing apparatus.

On the bright side, this gives him an excuse to clear the building of sightseers. Everybody who doesn't absolutely have to be here is gone in a flash. But it makes everything else more difficult, particularly the urgent business of communication. It's hard enough under normal circumstances to call somebody on the intercom, but when you're all wearing Scott Air Packs, it's like talking through a hose. It would be a good moment for panic if these men were of such a stripe, but Miller maintains a straight-ahead, business-as-usual exterior and everyone quickly falls in line. In the shift supervisor's office, they calmly pick up where they left off.

When Miller got here at seven o'clock, he found that Zewe and his crew were trying to cool the core with natural circulation; hot water rises and cold water falls, and this motion alone can be enough to carry away the heat. But when they discovered that the incoming core temperature had dropped to 200 degrees and the outgoing temperature was over 700, it was clear to them that the water wasn't moving.

Miller realized there must be steam in the system collecting in the high spots and blocking the flow. The only way to condense steam back into water is to drop the temperature or raise the pressure. Since they had no control over the temperature, Miller ordered Zewe to repressurize the reactor. The relief valves were closed and the pressure quickly shot up to 2,200 pounds. They have been holding it there by opening the block valve on top of the pressurizer every few minutes, but some of the engineers are starting to worry about how much punishment that valve can take. It was never designed for this kind of work, and right now it is their only way of controlling the pressure. If it fails, they are up Shit Creek.

More important, the strategy isn't working. They've been up to pressure now for two hours and there's still no sign the water's moving. The incoming and outgoing core temperatures are still 500 degrees apart.

From the beginning there has been a running "what if" faction in this discussion that keeps dredging up dragons of the deep. "What if the core isn't covered?" Miller feels they've got to address this possibility. Their voices muffled by the rubber face masks, they huddle between phone calls and try to figure out what to do next.

"Why don't we blow down all the way and float on the core flood tanks?"

Not a bad idea. The core flood tanks—part of the emergency core-cooling system—are the last bastion of defense in case of a major pipe break. These enormous tanks, each containing the equivalent contents of a swimming pool, are located just above the core. The water in these tanks is under pressure so that it will be literally blown into the system the instant the pressure in the core drops below 400 pounds per square inch.

What the engineers in this room have forgotten, or perhaps never noticed, is the odd piping arrangement that hooks the core flood tanks to the system. Similar to the pressurizer connection, the pipe forms a "U" like a sink drain. This "loop seal" means that the water in the tanks will not drain suddenly as they expect, but will flow in at a rate that depends on the pressure in the core. If they had a major pipe break, this would make no difference. The system would behave as designed, since the pressure would immediately drop to nothing. But with a slow blowdown through the relief valve, Miller will see only a trickle of water when the pressure drops to 400 pounds.

At eleven o'clock, Miller orders the men at the console to open the block valve once again and blow the thing down until the core flood tanks come in. It seems at last they have hit on a way out of this mess, but in fact they have only located another blind passage and they are steering ever closer to the reef.

And out in the dragon's lair, the cavernous bunker echoes again with the blast of venting gas. Only this gas is not steam, as Miller expects, but hydrogen, vast quantities of hydrogen created by the chemical destruction of the zirconium fuel rods. When hydrogen meets oxygen in the proper mixture it will recombine into H_2O with incredible force. The last historic recombination of hydrogen and oxygen resulted in the *Hindenberg* disaster.

11

From the air, the four towers are visible for fifty miles, and as the chopper roars west above the turnpike traffic, Jack Herbein strains for a view of his stricken plant. As Met Ed's vice-president/generation, he is responsible for almost everything but the light bills. He is also the senior technical man in the company.

Herbein was at a naval-reserve meeting in Philadelphia. He first heard about the accident when George Kunder called him from the control room at just about sunrise. It sounded as if they had things more or less in hand, so for the next couple of hours he sat in his hotel room and tried to deal with this thing by long distance. But at 9:30 he got a call from President Walter Creitz, who told him to get to the island right away; Creitz himself would arrange for a helicopter.

In a plant dominated by navy men, it is fitting that Jack Herbein is an Annapolis graduate. Like his deputy, Gary Miller, he has done time below decks as a ship's engineer, and like most of the crew on duty, he learned to run reactors courtesy of Admiral Hyman Rickover.

It's unusual to find a man with this kind of technical expertise this high up in company management. Most of Herbein's counterparts in other companies got there by way of the accounting department; they wouldn't know a pressurizer from a pump house. Herbein, however, actually holds a senior reactor operator's license. Slugging his way up through the ranks, he reached this lofty position the hard way, and he is a hard man. Some say he is a mean son of a bitch. But nobody doubts his competence.

As they approach Harrisburg, Herbein decides to set up shop at the Observation Center, a tourist information center that overlooks the island from the eastern shore. Herbein has faith in Gary Miller. They've worked together for a long time—they both watched the walls go up—and he knows that the last thing they need over on the island right now is a vice-president.

The pilot notifies Harrisburg approach control that he will be ducking in under the pattern to drop a passenger at Met Ed, and as the chopper hovers above the river, Herbein scrutinizes the wounded giant on the island. But the bleeding is internal. From here, it looks as powerful and grand as ever.

Across the river, however, there are plenty of signs of trouble. The Observation Center is under siege. It looks like a hundred people down there. They've got the place surrounded. As Herbein descends to the parking lot, they are all running toward him. The first contingent of the working press has arrived at Three Mile Island.

It is only a ripple in the tidal wave to follow. Even as Herbein fights his way through the microphones across the trampled lawn, more are arriving, unpacking their gear, long lenses aimed in all directions. He struggles to get inside the building, and a couple of company hard hats bar the door behind him.

"It's that goddamn Jane Fonda movie."

Indeed it was: one of those wild cards that can only be dealt by real life, a coincidence too outrageous for fiction.

It began in a Chicago saloon six years earlier. In the back room of Riccardo's restaurant one night in the fall of 1973, Dr. Henry Kendall let his spaghetti get cold as he sketched the outline of a nuclear reactor. Kendall, an M.I.T. professor whose work in high-energy physics is at the cutting edge of knowledge, was famous not only for his subatomic discoveries; he was also a founder of the Union of Concerned Scientists, a group of experts who were attacking nuclear safeguards on technical grounds.

Across the table sat Mike Gray, a bearded documentary-film maker who had a vague idea about writing a screenplay, and he was in the process of losing his appetite. Gray himself had a technical background—he studied aeronautical engineering at Purdue—and he was basically protechnology. Until recently, he was more or less pronuclear. But his research kept leading him to people like Kendall.

"If the fuel begins to melt," said Kendall, "it will become a solid white-hot mass of molten ceramic—highly volatile—and it will undergo chemical explosions as it comes in contact with other things."

Kendall is a slender, hawklike New Englander who comes from old money, which in part accounts for his independence. Most scientists are inextricably tied to government and industry for their grants and experiments. But if the bureaucrats tried to keep something out of Henry Kendall's hands, he could always write a check for it; his grandfather invented the Band-Aid.

"When the hydrogen buildup reaches 10 percent of the volume of the containment building," said Kendall, "it will explode spontaneously. All you need is one faulty weld to violate the containment. At Zion, 4 percent of the welding is faulty."

Stunned by Kendall's description of the ultimate accident, Gray scribbled in his notebook:

Gases initially can be lethal for 10 miles . . . inventory of strontium-90 enough to contaminate 10,000 sq mi . . .

In exposures of thirty minutes . . . 10,000 R, central nervous system would go in 24 hours . . . 1,000 to 3,000 R, vomiting, skin lesions, dead in a week . . ., minimum, 1.5 million people . . .

This was no long-haired environmental freak out to save the snail darter. This was one of the country's top physicists. And he was not chaining himself to the plant gate. He was attacking the design of the safety systems because the numbers didn't add up.

"When it melts through the bottom of the plant and hits ground water, the pressure will blow steam channels through the ground. Geysers of radioactive steam will erupt all around the plant."

It was late, and the maitre d' stood near the door waiting to clear the last table as Kendall delivered the punch line. "The emergency core-cooling system has never been tested," he said. "Our calculations show some situations where it won't work."

And if it did not work, and if Henry Kendall's worst fears were confirmed, the fallout could equal that of a thousand bombs of the size that fell on Hiroshima.

"We call it the 'China syndrome,'" said Kendall.

Over the next two years, Gray toured nuclear-power plants, studied blueprints, talked to engineers, dug through incident reports, and wrote. His screenplay was loosely based on his own experiences as a documentary cameraman.

In the fall of 1975, a copy of the third draft of *The China Syndrome* fell into the hands of one of the few Hollywood producers with both the power and the will to deal with heavy controversy. Michael Douglas, son of a superstar and a television hero in his own right, had just finished cutting *One Flew Over the Cuckoo's Nest*. Released just before Christmas, it swept the Academy Awards and marked Douglas as a major new force on the scene.

In the Goldwyn offices that once belonged to Howard Hughes, Douglas met with Gray in the spring of 1976. Douglas liked the story but he was cautious. "What about the technical stuff in this script?" he asked. "Is this true?"

"Afraid so," said Gray. "Almost everything in the script has already happened."

"What about this business where the needle sticks on the gauge?"

"That was at Dresden, just outside of Chicago, in 1972."

Like Gray, Douglas was obsessed with accuracy. The two men quickly came to an agreement, and the research and writing continued. At Gray's urging, Douglas hired as consultants three GE engineering executives who had resigned from the company's nuclear division because of safety concerns. At last, the filmmakers had the keys to the filing cabinet.

The first actor to call Douglas was Jack Lemmon. Then Jane Fonda saw the script and agreed to coproduce with Douglas. Fonda was, by any measure, one of the most popular actresses in the country. Critics might debate the fine points, but the accountants didn't have any doubts at all. Columbia pictures, sensing a blockbuster, pulled the stops.

On old Stage 4 at the Gower Street studios, Hollywood's incomparable carpenters built a precise duplicate of a nuclear control room, direct from industry blueprints. It was no illusion but a vast reconstruction that covered a quarter of an acre. An actor could enter the scene from the second-floor elevator, walk a hundred yards around the visitors' gallery, descend the concrete stairs to the security door, enter the control room past the working computer, step into Jack Lemmon's office, open his desk drawer, and pull out stationery with the plant letterhead. There was authentic trash in the wastebaskets.

It was Hollywood's first attempt to deal with heavy technology on its own terms. No scientist's daughter. No terrorist plot. Just good old American know-how gone wrong. Director Jim Bridges wove a web of memorable performances into an accident sequence that was researched by nuclear engineers and physicists. There was talk of Academy Awards for Lemmon and Fonda.

Partly because of her father's American Gothic image, partly because of her beauty, her outspoken independence, and her skill as an actress, Jane Fonda was by now a national institution. The press fought for interviews, and her picture graced the covers of a dozen magazines at once.

Armed with energy and charm, she and Douglas and Lemmon toured the country greeting every talk-show host and film critic from Amarillo to the Big Apple. They were preceded by a stark advertising campaign:

"THE CHINA SYNDROME"
Today only a handful of people know.
Soon everyone will know.

On March 16, the picture opened in 700 theaters, and all over the country there were lines around the block.

That was twelve days ago. This morning, what was intended as a warning of disaster has become a briefing film for the press. At a major newspaper the assignment editor waves the wire copy and shouts for attention. "How many of you guys have seen *The China Syndrome?*"

Three reporters raise their hands.

"You, you, and you. You're going to Harrisburg."

At the Capitol News Service just down the street from the state house they're doing a brisk business in instant nuclear research. Almost every magazine on the rack has a picture of Jane Fonda or Jack Lemmon, and the reporters are leafing through everything to see what, exactly, a "China syndrome" is.

On the back page of *Newsweek*, they find a blast at the film and everything it stands for. Castigating the producers for putting personal gain above the good of the country, columnist George Will says:

> . . . the sort of facts *The China Syndrome* deals with should restrict artistic license. They did not. All art is derivative from life, but a new genre is parasitic upon public affairs . . .
>
> To manipulate the audience into antinuclear hysteria, the film falsely suggests that nuclear-power companies carelessly risk destroying their billion-dollar investments.
>
> Nuclear plants, like color-TV sets, give off minute amounts of radiation, but there is more cancer risk in sitting next to a smoker than next to a nuclear plant . . .

All the same, Mayor Robert Reid would rather be sitting next to a smoker at the moment. His office in Middletown is less than 6,000 yards from the island, and so far all he's heard from Met Ed this morning is a lot of sirens. Reid, it turns out, is not on the list of people to be notified in a general emergency.

It's maddening. Every line to the island has been tied up since dawn, and it's the same with the company office in Reading. Finally, Reid manages to get through to Blain Fabian's office. What he hears is completely reassuring. Everything's under control. No radioactivity escaped. He feels as if a safe has been lifted from his back. He turns to the people standing by. "There's no problem," he says.

As Reid steps outside, he realizes how incredibly tense he

has become: fear of the unknown. In the absence of information, he let him imagination get out of hand. He gets into
his car and turns on the radio.

" . . . meanwhile at the capitol in Harrisburg, state officials
confirmed that there has been a release of radiation to the
atmosphere . . ."

To computer programmers, it is always shocking to recognize the similarities between the human brain and the
electronic variety. Like computers, we are programmed by
experience to accept some things and reject others. The
brain is wired to like Scotch or to hate it, and behind the
eyeballs, we each process data in a different way and give
half a dozen different accounts of the same car crash.

And if the information doesn't fit with our programming
—if it flies in the face of our experience—it is simply canceled. Nowhere is programming more evident than in the
mind of an advertising man, that specialized circuit board
wherein the client can do no wrong. Blain Fabian has the
mind of an adman. It's very hard for him to hear bad news.

At company headquarters in Reading, Fabian and the
public-affairs staff have gathered in his office to get an update from the horse's mouth. On the speaker phone is
George Kunder, the plant engineer, who is talking to them
from the control room.

"There's a possibility of core damage," says Kunder.
"Some external releases have occurred."

It is Fabian's job to put information in the best possible
light, and the best light he can put on this awful dialog is that
it must be speculation. Kunder, after all, is Miller's subordi-
nate. There's no reason to talk about core damage until they
get the word direct from Miller.

So the public statement is carefully crafted:

. . . At this time, there have been no recordings of any significant levels of radiation, and none are expected outside the plant
. . .

But as Pete Velez might remind us, radiation is like cow
shit; it's not so easy to keep it a secret from the folks downwind. Right now the wind is out of the south, and from the
slender steel stack above Unit Two, an invisible geyser of
radioactive gas is spreading into a broad plume, moving
gently up the Susquehanna toward the city of Harrisburg.

In the office at the State Bureau of Radiation Protection,
Tom Gerusky keeps a radiation monitor on the windowsill.
It's been sitting there for no particular reason since the
Chinese atomic fallout scare in 1976. Now it's starting to
click.

Click click.

Click.

Click click click.

Iodine-131 is one of the things you worry about. It is an
unstable cousin of the stuff on your medicine shelf. Your
thyroid uses traces of iodine in certain glandular transac-
tions, and it normally absorbs what it needs from things like
table salt. But the thyroid can't tell the difference between
normal iodine and its radioactive counterpart. When iodine-
131 "takes residence," as they say, it bombards the surrounding tissue with little packets of energy as the atoms fly apart.
If any of this subatomic shrapnel happens to nick the vitals
of a living cell, you have spawned a monster—a voracious
mutant that can multiply hysterically. It might not show up
for twenty years, but you have cancer.

Inside the core at Unit Two are a billion curies of iodine.
A significant fraction of this is now loose in the cooling
system, along with enormous quantities of radioactive

xenon and krypton gas. Nobody in the shift supervisor's office understands exactly how, but this deadly mixture is finding its way out of the containment into the auxiliary building. The floor on level 281 is under half a foot of water. In a desperate attempt to physically hold down the escaping gas, Gary Miller has ordered a crew to spread sheets of polyvinyl over the flooded area.

But gas has a way of getting around sheets of plastic. In the Unit Two control room everyone is wearing respirators, and over in Unit One they are reporting the bad news to NRC headquarters in Bethesda.

"I think I can talk to you before I get this mask on."

"Okay, Chuck."

"We are keeping up with radiation readings in Harrisburg."

"Have you got any idea of the magnitudes?"

"Okay, at Front and Market Streets in Harrisburg you are picking up one MR per hour. At the Harrisburg Mall, you are picking up five MR per hour. At the turnpike exits they are picking up twenty-five MR per hour."

"Dispatch somebody immediately to check the Harrisburg with measuring stuff to start taking confirmatory measurements. Send somebody down there immediately. Okay?"

"Hold on."

"I guess we don't believe it."

"This is uncalibrated."

"Yes, we heard that."

"I'm going to have to go on a respirator, so hold on and we'll see if I can talk to you, okay?"

The Pennsylvania state house has a feeling of ornate gravity, and there is a good reason for this. In a deal that makes modern-day legislators palpitate with envy, the contractors who built this edifice managed to sell the fixtures and furniture to the state at a fixed price per pound. So it is, in a sense, one of the heaviest state capitols ever built. Even in the smallest offices, the chandeliers weigh hundreds of pounds, and it would take a forklift to rearrange the furniture. The apex of this opulence is the third-floor corner suite that houses the governor.

Seated behind a desk that makes the Oval Office look like a rented set, Dick Thornburgh is watched over by oil portraits of the past statesmen and con artists who have occupied this room, and this morning they stare down at him with icy indifference to his awful dilemma. Thornburgh is the man who must decide whether or not to start moving people. All things considered, this Wagnerian setting strikes him as appropriate.

Evacuation would be easy. It's been on his mind ever since Colonel Henderson called him away from breakfast. Just hit the switch and say, "Everybody out!" But evacuations are always full of surprises, and they aren't cheap. There's a Bethlehem Steel plant, for example, not two miles upriver from Three Mile Island. You don't just shut down a steel mill for the hell of it. Also, you can kill people. The most recent evacuation was in Florida, and six people died in traffic during a routine retreat from a hurricane.

This, however, is no hurricane. This is the unseen unknown. Nobody's ever dealt with this one before. In Thornburgh's mind is the thought of a very real possibility that things could get out of hand. You can see a flood coming and you know when it's gone. But this? One slip and he could start a panic that would go down in history.

Over the past hour, Lieutenant Governor Scranton has managed to collect an entourage of technical people from the various state agencies and has herded everybody into the governor's office to tell him what they know. Some, like Colonel Henderson, are specialists in evacuation; others know something about radiation—inevitably there are one or two lawyers in the crowd; and Bill Dornsife from the Department of Environmental Resources is an honest-to-god nuclear engineer. What's more, he used to work for a company that did some consulting for Met Ed, so he's actually laid eyes on the machine in question.

This is what Thornburgh needs. Facts. First-hand impressions. The best judgment in the world isn't worth a damn without facts. Dornsife does his best to give the governor an instant education. What he has to say is basically reassuring. There has been a venting of radiation to the environment but there is nothing substantial off-site. There's no threat to public health. They're still trying to figure out exactly what happened, but it sounds like they've got everything pretty well in hand.

A quick check around the room shows they are all of one mind. There's no indication at this point that anybody's in danger. The governor tells them to get on top of it and get back to him as soon as they have anything more substantial.

But the meeting ends on an unsettling note. Outside is a gaggle of reporters waiting to see Scranton. They want a clarification. It seems there is a conflict between what Scranton has been saying and what they're getting from Met Ed.

"Sir, in your press conference you said there had been a release of radiation from the plant . . ."

"That's right," says Scranton. "Our understanding is they had a small release of radiation but there is no danger off-site."

"A company spokesman in Reading just said there has not been a release. That it was strictly on-site."

Good God, now what? It's bad enough that nobody down at Three Mile Island is able to speak plain English. Now they can't even get their stories straight. Scranton heads for his office in the other wing of the capitol. He doesn't like the feel of this.

When he gets there, one of his aides intercepts him. "We

just got a call from the Bureau of Radiation Protection. They say Met Ed has been venting radioactive steam directly to the atmosphere since about eleven o'clock."

Like a bolt of lightning, this news ricochets through the executive offices. Scranton is outraged. At the very moment when he was on camera telling the press that no more releases were expected, Met Ed was in the middle of another release. He tells one of his aides to get Met Ed on the line. He wants somebody up here from the plant who knows what the hell is going on, and he wants to see them here, in his office, now.

The pressure on Gary Miller is hard to imagine. From the minute he came aboard he's never had less than one phone in his hand. On top of everything else, he's had to spend a great deal of time explaining nuclear power to every big shot who managed to get through to the control room. Naturally, everybody wants to talk to the head man.

And for the past hour, he's also had an open line to his boss, Jack Herbein, over in the Observation Center. Herbein is adding to the pressure. He wants Miller to stop venting steam.

The argument that erupts is loud enough for everybody in the control room to feel the heat from across the river. But Miller is venting steam to the atmosphere for a very good reason. At the beginning of this whole affair, back in the wee hours of the morning, the initial jolt from the water hammer fractured several control lines down in the turbine building; that put the main condenser out of business. Since they can't dump steam into the condenser, where it would normally go, they've been dumping it into the atmosphere. But Herbein is afraid this steam is the source of the leaking radiation.

Miller reminds him that this is the only heat sink he's got right now. If the core is getting any cooling at all, he believes, it's coming from the water that's circulating through the boilers. Water from the core passes through the boilers, where it gives up heat to pure water on the other side of the pipes. The steam from these boilers will be clean unless there is a break in the pipes.

Miller acknowledges that pipes in one of the boilers have cracked—apparently fractured in the opening moments of the accident when the boiler dried out—but they've had that loop sealed off for hours. He is absolutely certain that the steam he's venting from the other boiler is not the source of the radiation they are seeing.

"We sent a guy up on the roof to measure the activity," says Miller. "He got nothing."

All the same, it doesn't look pretty. Steam is visible, and that cloud rising over the island is giving everybody the jitters. Miller feels the pressure inside, but Herbein is feeling the pressure on the outside. The building he's standing in is surrounded by reporters, and every little old lady along Highway 141 is probably looking out the front window at that goddamn steam. Herbein wants it stopped.

Like the captain of a ship, Miller is in charge of the plant; there is no question that his word is final until he's relieved. On the other hand, Herbein is his boss, and he's a hard man to ignore. Reluctantly, Miller agrees to cut it off. He sends word out to the control room to shut the atmospheric dump. They'll have to figure out something else.

Throughout this exchange, Miller is wearing breathing apparatus. And this brings Herbein to his next point. Since the control room is filled with radioactivity, it stands to reason that some of it must be leaking out through the ventilation system. Herbein suggests that they shut down the control-room ventilation. Miller argues with him—his language shows the strain of six hours in heavy weather—but Herbein can be persuasive. And he prevails.

His motives are the best: Herbein wants to stop the leak

wherever it is and confine the radioactivity to the site. But he came into the game late, and his perspective is filtered through the phone. He can look out the window and see the plant, but he has only the most general picture of what's going on inside. Miller is not much better off, even though he can look out the window of the shift supervisor's office and see the control room. The giant machine these two men have nurtured into existence has suddenly developed a mind of its own.

Almost as soon as Miller shuts down the ventilation system, the radiation levels start building. In the control room, the operators, the engineers, and the chemists, like spacemen in their breathing masks, all have an eye on the rising numbers. At this rate, it won't be long before they have to run for it. Miller orders the ventilation system back on, and he decides not to bother Herbein with the details.

At the moment, however, Herbein has some problems of his own. Company president Walter Creitz is on the other line. Creitz just had his ass blistered by the lieutenant governor. He wants Herbein to get up to the capitol right away.

This is exactly what Herbein was afraid of. God damn it. Venting that steam was a bad idea. He gets back on the line with Miller. As long as he's going to face the lions, no sense in going alone. He tells Miller that he wants him to come along.

Miller is stupefied. It will take an hour just to get up to Harrisburg and back. It's like leaving the bridge in the middle of the battle! Agreed, he's had to spend most of his time on the fucking telephone, but he's totally involved in everything that's going on in the plant, and they still aren't sure exactly what they've got here. To leave now is out of the question.

Herbein is adamant. The people up in Harrisburg want a face-to-face meeting with somebody who has first-hand knowledge. He orders Miller to meet him at the Observation

Center as soon as possible. Herbein hangs up. Then he looks out the window at the sea of reporters covering the lawn.

Amazing. Camera crews all over the place unpacking from vans, transmitter trucks jockeying for position. There's ABC . . . CBS . . . Jesus, what a mess.

Obviously, somebody's got to talk to them. The only PR man in sight is Bill Gross, an English teacher who works part-time as a tour guide. Herbein has never done anything like this before. But he's not the kind of guy to duck an ugly assignment. He grits his teeth and steps outside. Bill Gross shouts for attention. He announces that Vice-President Jack Herbein will answer questions, and the mob surges forward.

It's an unfortunate conjunction. Herbein is an engineer and a good one. But like many engineers, he's led an obsessive life that has given him great experience with an increasingly narrow field of view. Reporters are a mystery to him. He doesn't understand their ground rules, or even exactly what they're up to. And like most engineers, he doesn't pay much attention to clothes. To a man like this, a good suit is one with the right number of pockets. Unfortunately, he's in the middle of a divorce, and the suit he packed for his trip to Philadelphia is a black-silk number that gives him all the polish of a Chicago alderman.

On top of this, he talks an unfamiliar language. Every other word is something nobody's ever heard of. A couple of network cameramen notice in their viewfinders a disturbing similarity between Jack Herbein and his movie counterpart in *The China Syndrome*. Using almost identical words, the spokesman in the movie says, "There is no danger to the public . . . ," and he is covering up.

Gary Miller is making preparations to leave the island. He will transfer command to Joe Logan. He tells George Kunder, the plant engineer, to collect all the data he can find; he wants Kunder to come with him to Harrisburg. Like his boss, Miller wants some company at this meeting.

It's insane for him to have to make this trip, but if he has to go, this is as good a time as any. The situation looks fairly stable. They finally managed to get the pressure down to a point where the core flood tanks released, and very little water drained into the core. To Miller and the rest of the crew, this says the core must be covered.

Just then a slight shudder ripples through the building; in the control room, Miller and the others hear a distant "thud."

"What the hell was that?"

"Must have been the ventilation dampers."

But Bill Zewe and a couple of supervisors near the console catch an ominous signal. One of the recording instruments —a pen that traces pressure in the containment building— takes a sudden leap.

"Look at that."

In an instant, the needle bounces a third of the way up the scale. Astonished, they watch as the line peaks at twenty-eight pounds per square inch, then slowly descends.

It's a number that's impossible to believe. The containment building is an enormous empty room nearly thirteen stories tall. To create that kind of a pressure wave in such an immense space would take a phenomenal explosion. The instrument must have failed. Probably a stray electrical current.

But a couple of the other guys on the console are not so sure. What about that "thud"?

This discussion is going on as Miller and Kunder walk through the control room, heading for the door. But their minds are already racing ahead to the meeting with the governor. And perhaps it's a piece of information they simply don't want to know. One last look around and they are gone.

Had he paid any attention to the readings, Miller could have made a quick calculation and realized that a cloud of hydrogen had just exploded inside the containment building with a force of 200 thousand tons.

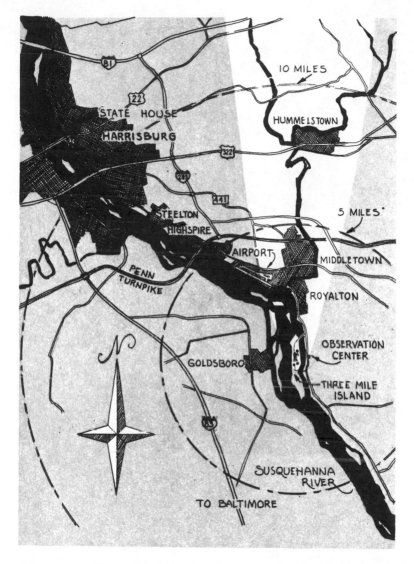

Figure 7. The radioactive plume at 4:30 P.M., on Wednesday, March 28. At a distance of 16 miles from the stack, readings drop off to 0.1 millirem per hour. It is assumed to be primarily xenon–133.

12

Of all the people who foresaw this awful moment, none was more vocal than Jim Creswell, the Chicago area inspector who hand carried his fears all the way to Washington a scant six days ago. Since he got back to the office, he's been rubbing his neck and waiting for the ax. Everybody here knows what he was up to. It's just a question of time until they make life so unbearable he'll have to get out of here.

This morning he came in bright and early; no sense giving them an excuse. They've been piling work on him—busywork in triplicate—to keep him out of the hair of the people in Toledo. At the moment, however, he finds himself staring at the corner of the ceiling, trying to figure out his next move.

A couple of the other inspectors are secretly pulling for him. They're as worried as he is about the casual, ragtag handling of safety concerns within the agency. But they keep it to themselves. They wouldn't be in his shoes for all the gin in Bombay.

Just then one of these silent partners sticks his head into the office. "You heard? There's been an accident in Pennsylvania."

Creswell rocks forward and clicks on the radio.

". . . which began early this morning when a valve failed and caused the reactor to shut down automatically. State officials say there was a minor release of radiation but there is no need for evacuation . . ."

A couple of other faces appear in the doorway. Creswell looks up. "I think this just saved my ass."

Like everyone else, he assumes the worst is over.

The one guy they could have used in the control room this morning is Jim Floyd, a man who knows this reactor so well that he might have been able to get an early grip on the situation and turn it around. Unlike the others, Floyd has a feel for the machine that you can't get from a book. He knows how to fly that baby, and he doesn't need a chart to tell him that hot water boils when you lose pressure. But the luck of the draw finds him five hundred miles away in Lynchburg, Virginia. Jim Floyd is the operations supervisor, and he's down here with the A crew, some of the most experienced operators from Unit Two. Together, these men probably know the plant better than anyone else. Unfortunately it's their turn to train on the reactor simulator at the Babcock & Wilcox factory.

Floyd heard about the accident when he came down to breakfast at the Sheraton coffee shop. One of the guys in the crew is dating a guard at the plant; she called to tell him something was going on at the island. She said the main safeties had been blowing for two hours.

This excited Floyd's imagination. He knocked back his coffee and they all left for the Babcock & Wilcox plant. Floyd had an unlisted number for Three Mile Island, and as soon as he got through to the control room, he realized they were in shallow water. On the basis of the radiation numbers alone, he could tell off the top of his head that at least an eighth of the fuel pins had already failed.

He told them he would try to duplicate the accident on the simulator and see if he could figure out from here what was happening. Working with the Babcock & Wilcox instructors, he programmed the data into the simulator and fired it up. But no matter which way they played it, they weren't able to imitate what was happening on the island.

This stands to reason. If the men in the control room had recognized the essential fact—that the cooling system had had an uncontrolled leak for nearly two and a half hours—they wouldn't have needed Jim Floyd to solve the problem on the simulator. Floyd has no knowlege of the business about the block valve on top of the pressurizer, so each time he runs the sequence he draws a blank.

Also, the reactor simulator itself has a couple of shortcomings. For one thing, it isn't programmed to deal with an accident like this, since nobody in the training department thought it was possible. And the simulator panel doesn't vaguely resemble the control room at Three Mile Island. While it has all the basic controls, the hundreds of alarm signals and corridors of extra instruments are missing. If Floyd had been on the island in the opening seconds of the accident to see the console turn into a jangling Christmas tree, he might have suspected that some of the details were overlooked.

An hour or so later, Floyd calls the plant and gets an update from one of the supervisors in the control room. The new information is punched into the simulator and he does his best to bring the damn thing to its knees, but he can't drag the pressure down anywhere near 1,000 pounds. They are in the middle of another run when Floyd gets a call from upstairs; some of the management people would like to see him in the executive conference room.

A guy like Jim Floyd doesn't normally travel in these circles. When he gets up to Mahogany Row, he finds about

forty grim-faced engineers he's never seen before. These are
the top brains in the company, the men who actually designed the reactor at Three Mile Island. Floyd has something they desperately need: the unlisted phone number for
the control room.

Babcock & Wilcox was notified at about 7:45 by Leland
Rogers, their man on the island; they told Rogers to find out
what he could and call back at nine o'clock. Then they
quickly assembled the engineering staff. But the call never
came. They've been trying since to reach the plant, but all
they can get is a busy signal.

Like Floyd, these men have only snippets of information,
most of it third-hand, and none of it sounds good. There is
a broad spread between the core inlet and outlet temperatures, they are having trouble with the pumps, and the emergency core-cooling system was triggered automatically. For
some reason no one here completely understands, the operators apparently throttled back on high-pressure injection.

Across the table sits a man who is completely baffled by
this news. As chief designer of the emergency core-cooling
system, Bert Dunn is under the impression that a memo
went out last year warning the operators not to cut off
high-pressure injection in the middle of an accident. Yet
here they are, backing off the emergency systems, just like
at Davis-Besse a couple of years ago. Is it possible . . .?

Dunn and the others have come to the conclusion that
whatever the hell else may be going on up there, they've got
to get high-pressure injection flowing to the core. That's
why they sent for Floyd. They are hoping he can get a
message in to the control room.

There is a squawk box at the head of the table. Floyd pulls
out his wallet, checks the secret number, and dials.

Busy.

Okay, he's got another number. Not as direct, but at least
he's able to get through to the control room over in Unit
One.

"Unit One. Hutchinson."

"Rich? This is Jim Floyd. I'm down in Lynchburg. The people here at B & W have some information, and they say it's very important to get this to the foreman in Unit Two."

One of the engineers at the table has scratched some notes on the back of an envelope. Floyd reads it. "They say the high-pressure-injection flow should not be less than 400 gallons a minute. Five hundred would be better. Send a runner over and get ahold of the foreman and tell him he has got to keep the HPI above 400 gallons per minute to remove the decay heat. Make sure that word is put into Mike Ross's ear. Mike Ross has to know this. It is very important."

As the machine's designers hunch over the table, straining to hear the squawk box, the voice from the island promises to get word to the Unit Two control room. The sense of relief around the conference room is interrupted by another call.

Leland Rogers is only one of many Babcock & Wilcox employees permanently based at Three Mile Island. Another engineer, Greg Schaedel, was home in bed when the accident started. He's been trying since then to get onto the island, but the guards won't let him across the bridge. So, like Jack Herbein, Schaedel finally wound up at the Observation Center, where he could at least talk to the control room on the company line.

Schaedel has the latest figures from Leland Rogers, and he's calling headquarters to relay the information. Now, over the squawk box, Floyd hears for the first time that the main coolant pumps have been off since six o'clock this morning.

Floyd rocks back in his chair. "Those boys are in trouble," he says.

Schaedel's view, however, is strictly upbeat. He says the staff thinks they have collapsed the steam bubble. But as he reads off the temperatures and pressures, Floyd sits bolt upright.

"No way is that liquid!" he shouts.

As the astonished vice-presidents listen open-mouthed, Floyd yells at Schaedel, "They have not collapsed the bubble in the hot legs yet with those numbers! They damn well ought to look at their steam tables! The bubbles have not collapsed as long as those hot-leg thermocouples are above saturation temperatures for the pressure they are at."

Floyd doesn't have a set of steam tables with him, but he knows instinctively that the temperatures are way over the line. It's incredible. There's Schaedel, and Leland Rogers, and a plant full of graduate engineers, and nobody seems to remember that water boils at any pressure if you get it hot enough.

Down at the other end of the table, Bert Dunn does have a set of steam tables. He quickly confirms Floyd's educated guess. The outgoing core temperature is well above the line. There is only one way that can happen. Steam is forming in the system and is being superheated by something that is hotter still.

The core is uncovered.

Now Dunn and Floyd are both yelling at Schaedel on the speaker phone. They tell him he's got to get word through to the control room to crank up the high-pressure injection to a minimum of 400 gallons a minute.

At the Observation Center, Greg Schaedel hangs up, shaken to his boots. He's just had his ignorance of fundamental steam physics exposed to company management, but that's the least of his problems. As he picks up the line for the control room, he looks out the window at the gray fortress across the river.

Mother of Jesus.

Despite all the plans and lists and charts and drills, the Nuclear Regulatory Commission is ill-prepared to handle a crisis, and the problem can be traced right to the top. The agency is run by five independent, coequal commissioners; not the ideal setup for making hasty decisions.

Of the five members, only Chairman Joseph Hendrie has practical experience in nuclear power—he came with the agency when it was lopped off the old AEC—but Hendrie is at a hospital in Virginia, where his daughter is undergoing dental surgery.

Commissioner Richard Kennedy is in today. But Kennedy seems less concerned about the health of the stricken reactor than about its image. The public-relations office is about to put out a press release, and they've brought it to the commissioner for approval. Kennedy doesn't like it.

"One thought I had—I had two thoughts," says Kennedy. "You use the word 'accident' twice."

"Yes, sir."

" '. . . considered an accident . . .' People think of accidents, you know . . . in the context of *The China Syndrome.* Is this an accident?"

"I believe it's an accident, Mr. Kennedy."

"Define an accident."

As the car heads north along the river racing for the capital, Gary Miller and George Kunder try to tell Jack Herbein everything they know, in the minutes before they meet the lieutenant governor. Miller carries a phone beeper, and his car has a company radio so that he doesn't feel completely cut off from the island. All the same, it's probably a good thing people along the highway don't recognize them. If they knew that the plant's top technical staff was leaving the scene at high speed they might get the wrong idea.

Like the Observation Center, the capitol building is

swarming with reporters. Trying to attract as little attention as possible, the three engineers hurry up the sweeping marble steps of the rotunda, past the regimental flags once carried against the rebels at Gettysburg. One of the lieutenant governor's aides is on the lookout and hustles them into Scranton's office.

The room is packed. In addition to the traveling band of experts Scranton picked up at the press conference, he has added an attorney from the justice department, just in case they have to start bending arms. There is a quick round of introductions but no small talk.

"We are very concerned that you've been going ahead and venting this thing and not telling anybody about it," says Scranton.

Miller tries to explain that the steam he was venting had only the most minimal amount of radioactivity—barely detectable—and well below any emergency-action level.

Scranton doesn't give a damn. "Whether or not it's dangerous, this is not the way to proceed in an emergency situation." He tries to explain to these technocrats that there is a very real need for the folks in the street to understand what's going on and to get it in their own terms.

Herbein tries to calm everyone down. He downplays the incident. He doesn't actually say the reactor is under control, but he makes it clear there's nothing to worry about.

Scranton is annoyed. If there's one thing he doesn't need right now, it's some guy with a leave-it-to-us professional knowingness. After listening to some parabolic explanation, he asks Herbein point-blank, "Have you been venting radiation?"

"Yes," says Herbein, "but the off-site releases were minimal."

"Did you tell the press this release had been made?"

"They didn't ask me," says Herbein.

Like a tree falling in the empty forest, Herbein's credibility comes crashing silently to earth. He is too caught up in his own thoughts to notice, but an icy wave of apprehension sweeps the room. It's as if a rattler had slithered out from under the desk. Here is the chief spokesman for the company, and it turns out the son of a bitch won't give you a straight answer unless you pin him down.

Standing now, the lieutenant governor warns Herbein not to release anything to the atmosphere—or the river— without letting the state know in advance. Then he leaves for the governor's office to let Thornburgh in on the bad news.

Neither of these two political leaders has any specific credentials for dealing with this awesome moment. Neither has any special knowledge of radiation, or nuclear power, or evacuation. But they share one insight that stands between the citizens of Pennsylvania and chaos. This office, and the man behind this great, godawful desk, must maintain credibility at all costs. Thornburgh and Scranton must tell everything they know as soon as they know it—no hesitation, no equivocation, just the plain simple truth—or they will be writing a scenario for disaster.

At 4:30, Scranton enters the packed press room for the second time today. The atmosphere is crackling with rumors. Frozen in the light of the flashbulbs, he steps to the microphones. It's clear from his opening sentence that the deal has changed.

"This situation is more complex than the company first led us to believe," he says. "We are taking more tests. And at this point, we believe there is still no danger to public health. Metropolitan Edison has given you and us conflicting information. There has been a release of radioactivity into the environment. The magnitude of the release is still being determined, but there is no evidence yet that it has resulted in dangerous levels. The company has informed us that from about eleven A.M. until about 1:30 P.M., Three Mile

Island discharged into the air steam that contained detectable amounts of radiation . . ."

Around the base of the Washington Monument, the flags snap smartly in the March breeze as black limousines glide by on Constitution Avenue. Up on the Hill, concern about the situation in Pennsylvania is beginning to filter into the halls of government.

The White House was notified at 10:30 A.M. when a cable came into the situation room and was handed to Jessica Tuchman Matthews, an aide to Zbigniew Brzezinski. Brzezinski, ever the eager point man on any issue, snatched up the cable and took it immediately to the Oval Office.

Now, senators and congressmen are calling friends and connections at the NRC trying to get some kind of handle on this thing. What they are hearing on the radio doesn't sound all that terrific. The NRC staff up in Bethesda is reassuring, but it turns out they actually have little in the way of substantial information. Even though the accident has been in progress now for over twelve hours, it seems the NRC has still not been able to set up a direct line between Bethesda and Unit Two. All those numbers and technical questions must be relayed either through the regional office outside Philadelphia or through the operators in Unit One. And at the end of the line is some guy in a respirator who has to send a runner from one control room to the other.

But though the data are fragmentary and the numbers are garbled, it's all starting to make an impression on Victor Stello. Director of inspection and enforcement, NRC headquarters, Stello is a big man. Immense, even. Tipping the scales at 250 pounds, he has the bulk and reputation of a bulldozer. His enemies see him as a bully who has managed over the years to ramrod policies and regulations through the agency that suit the nuclear-power industry. His supporters say that without the aggressive leadership of a man like Stello, the agency would never get anything done. Respected by friend and foe alike, he has plenty of each.

Stello came with the organization when it was severed from the old Atomic Energy Commission. He is one of the top technical minds on the headquarters staff, and he's been at the Incident Response Center in Bethesda for most of the day.

For a while now, Stello and some of the others at headquarters have been aware of the anxious vibrations coming out of Lynchburg. The Babcock & Wilcox engineers have called several people to suggest the grim possibility that the core might not be covered. Stello decides it's time to get a fix on some of the numbers to see if there's any substance to this.

He has come into the communications room to get a little closer to the source. Right now the man at the switchboard is in contact with a supervisor in Unit One; the latest numbers are just coming in.

"We were asking—the discrepancy between the 450 PSI and the 500-and-whatever-it-was temperature . . ."

"Five hundred fifty," says the voice from the speaker.

"The 550 temperature. This was T-hot, which would have said you had some superheat there."

"Yeah, well, what's the question?"

"Well, we were wondering if that is a valid temperature."

"Well, I don't—you know, I don't know how accurate those temperature indicators are up there, because at one time T-hot was pegged high at 620, right? . . . and I guess T-cold was what, something about 220 or something?"

"Yeah, 220—200 was the latest we got, I think."

"I talked to somebody about that and I said I wouldn't put much stock in those T-hot instruments. We don't know whether they're true or not."

Standing here listening to this, Vic Stello can't contain

himself any longer. He leans over the desk and grabs the phone.

"Let me bounce a question off you," says Stello. "If you really have 550 degrees on that hot leg, it's clear that you're getting some superheat. If you're getting superheat, there's a chance the core could be uncovered . . . Have you thought about what problem you've got, if indeed you've got 550 degrees on that leg at 450 pounds?"

The distant voice is thoughtful. "Yeah, I see what you're saying . . . They do have . . . 175 inches indicated in the pressurizer, which means that the core would be covered. They've also got the core flood tanks floating on that."

Stello is exasperated. "But that doesn't necessarily mean that they don't have a steam bubble in there."

"Oh, okay, you're talking about a steam bubble in the core."

"Yeah," says Stello, "and if you have a steam bubble in the core you've got the top part of the core—which could be uncovered—superheating the stuff coming out of there, and that's what's giving you the reading. Have any of you people out there talked to B & W about what kind of a problem that could look like?"

"I don't know if they've talked to them over in Unit Two or not, sir."

Stello says, "Well, I know they've been trying to get in touch with you, as we best can understand it, and as we're talking to them we see the same concern and the same problem. If that thermocouple reading is correct and you do have superheat coming up through the core, there's a question——"

"I'll talk to them. Okay?"

"Huh?"

"Let me go, let me talk to the guy at Unit Two and see what they have—you know——"

But Stello has a second thought. He doesn't want to put the NRC on record as making a specific recommendation. If Met Ed took the action he recommended and it backfired, the NRC could be responsible. He wants the man to understand that Met Ed is still in charge of the accident.

"I'm not telling you to," says Stello. "I'm asking you to consider it."

"Huh?"

"I'm not telling you to. I'm asking you to consider it."

"I understand . . ."

But suddenly the idea has lost a little urgency. It's no longer a command from a high federal official. It's just a suggestion. And the guys in Unit Two have been flooded with suggestions all day. This one goes on the list.

It's about an hour before sunset when Gary Miller crosses the bridge on his way back to the island. The trip to Harrisburg took even longer than he feared, and after dropping his boss back at the Observation Center, he and Kunder are hotfooting it over to the control room.

There is one bright note. In their absence the radioactivity in the building has dropped to a point where the operators have been able to take off the cumbersome breathing apparatus. Apart from that, the situation is about the way they left it.

Almost as soon as he reaches the shift supervisor's office, Miller finds himself back on the line with the Observation Center across the river. Once again, Jack Herbein is getting heat from the outside world. This time it's coming from top management. They've been in contact with Babcock & Wilcox, and everybody down in Lynchburg is concerned about the core. They're afraid it might not be covered. They say it is imperative that Miller get the pressure back up to collapse the steam out of the system. Also, they want him to keep high-pressure injection going to carry away the decay heat.

Miller objects. He believes the core is covered. He thinks the way out of this mess lies in the direction of depressurizing the system. He wants to let it blow down as low as it will go and then start a normal shutdown procedure. Herbein, however, is impressed by the weight of all that brass on the other side of the argument. This is no hot tip from the janitor: this is a message from one boardroom to another. He orders Miller to shut the block valves and repressurize the reactor.

Finally, five hours after Babcock & Wilcox management first began frantically trying to get word to the control room, somebody in authority has direct orders to turn on the emergency core-cooling system.

Miller knuckles under. After a quick huddle with Kunder and the others, word goes out to the men on the console. "Shut the block valves and repressurize. We're going back up to 2,100 PSI. Take the locks off the ECCS and maintain HPI at 450 gallons a minute."

Paula Kinney can see the cooling towers from her kitchen window, but she's never given it much thought. They are fixtures. Part of the landscape. But her mother called just as she was putting lunch on the stove and told her to get the kids and run for it. It was the first she heard that anything was going on.

"Oh, Mom, I don't think there's any problem. If it was serious they'd let us know."

She promised to turn on the radio and stay in touch. But throughout the day, the local news played everything pretty low-key and Paula felt sure that nothing very important was going on. So it is something of a shock when she

flicks on television to catch Walter Cronkite before din-
ner.

> "Good evening. It was the first step in a nuclear nightmare; as
> far as we know at this hour, no worse than that. But a govern-
> ment official said that a breakdown in an atomic-power plant
> in Pennsylvania today is probably the worst nuclear accident
> to date . . ."

The main cooling pumps, those four immense thousand-
horsepower impellers that normally carry away the reactor's
heat, have been shut down since early this morning. Almost
every nuclear engineer who hears this feels as if there is a
gun in his face, and the plant's designers are no exception.
The staff in Lynchburg says it is essential to get one of those
pumps running again.

By 6:30 P.M. a phone line is finally established between the
Babcock & Wilcox plant and the Unit Two control room.
It has taken most of the day, but at last the men who built
the reactor are able to speak directly to the men who are
running it.

Lynchburg says the most urgent priority is to get water
moving through the core again. But it turns out there is a
whole list of obstacles to starting the main pumps. For one
thing, there has been an electrical failure down in the auxiliary building, and the system that supplies oil to the pumps
is out of business. They don't dare start the pumps without
oil. But the shorted panel is in the middle of a high-radiation
area. Somebody has to go down there and wire around it.

You don't send a lone operator on a mission like this. Two
men must go: one to do the job and another to stand clear
and watch. Otherwise a man might slip and fall and be dead
before anyone knew he was missing. A couple of men are
carefully suited up while Miller and Dick Dubiel personally

supervise. Two layers of gloves, coveralls, a double wet suit, inner boots, outer boots, Scott Air Pack—a process that takes nearly twenty minutes. The operator is briefed on the job. Looking over the blueprints, he rehearses it in his mind. Then they are off.

As the door closes behind them, Miller and the others return to the conference call with Lynchburg. One of the other problems they have encountered is the safety system that protects the main pumps from self-destruction. Following detailed instructions from the men at the conference table 300 miles away, the plant operators figure out how to get around the computer logic and put jumpers on the circuit.

A few minutes later, word comes back from the auxiliary building. The oil-lift pumps are operational.

Everything is ready, but nobody is particularly eager to hit the switch. This is, after all, a first—a practical experiment on a live reactor. Nobody is sure what will happen.

Lynchburg suggests a trial run—a short "bump" to see how it goes. Feeling more than a little suspense, Miller gives the signal at 7:33 to close the contact on reactor pump 1-A. As a dozen pair of eyes race over the instruments, he lets it run for ten seconds, then shuts it down.

All the indications are good. The temperature dropped instantly and the vibration was bearable. Lynchburg gives the go-ahead. The operators cross their fingers and start the pump, and for the first time in fifteen hours water is once again moving past the core.

There are a lot of people in bed tonight who aren't sleeping, and one of the most fitful of them is in the governor's mansion. Dick Thornburgh stares at the sculptured ceiling and replays the day in his mind. But he keeps hanging up on one point. In all the briefings—Scranton, Met Ed, the

NRC—nobody ever mentioned the one thing that was surely in the back of everybody's mind.

What about the core?

There is probably a good reason they've all been dancing around this question. They are afraid of the answer. Like a trained dragon that has lost its fear of the whip, the 150-ton core is no longer able to perform tricks, but it is dangerous beyond imagination.

All 36,000 fuel pins have burst. All of the radioactive gas sealed in these slender tubes has escaped into the system. Over two-thirds of the core has been uncovered for periods as long as two hours. The fuel itself has not actually melted yet. But in the upper few feet, it has formed a chemical solution with the molten zirconium, and this white-hot mass has dripped like candle wax into the water below. Then it fragmented and collapsed in a pile of rubble that is trapping steam in pockets deep within the fuel bundles. The melting point of uranium—the point beyond which nothing can save it—is 5,000 degrees. Parts of the core have reached 4,300 degrees.

13

Thursday, March 29

The highway centerline dashes out of the blackness, flashing past like tracer bullets as Jim Floyd rockets through the night bound for Harrisburg. Floyd, the Unit Two operations supervisor, was in training down at Babcock & Wilcox when this whole thing began, and he's been anxious to get back to Harrisburg ever since.

It's raining on Interstate 81, but Floyd doesn't need the sound of the wipers to keep him awake. He suspects the plant is in deep trouble, even though the occasional news broadcast he manages to pick up makes it sound as if the thing is over. It's a gut feeling. From the moment he heard about that first phone call at breakfast, he's been cracking his knuckles. And that scene in the boardroom was something else. All the top honchos at Babcock & Wilcox in one room, waiting for Jim Floyd. And what did they want? His secret phone number for Unit Two.

That session went on for a couple of hours while they tried to get through to the control room. Then one thing led to another, and he wound up stuck in Lynchburg until nearly nine o'clock. In this rain, he'll be lucky to reach Harrisburg before two or three A.M., but he's made up his mind to go straight to the island.

Jim Floyd's boss's boss's boss, however, doesn't share

Floyd's anxiety about the plant. Up in Harrisburg, Met Ed president Walter Creitz is more concerned about the incredible damage that is being done to his company—and to him personally—by the misinformed reporters who are descending on the plant like locusts. Creitz and his deputy, Jack Herbein, agree to split up the morning television interviews. Herbein will take "Thursday Morning," on CBS, and Creitz will handle ABC and the "Today" show.

At five A.M., exhausted and defensive, Creitz shows up at the WTPA–TV studio in Harrisburg to tape a segment for "Good Morning America." For balance, the ABC producers have nuclear critic Dan Ford standing by in Washington. Ford, along with Henry Kendall, founded the Union of Concerned Scientists, and what these scientists are concerned about right now is the fuzzy reports coming from Creitz and his people. Ford is blunt and to the point. "If the equipment disablement isn't clear to you yet, how are you certain that the plant is now safely under control?"

Creitz says, "Well, the radi—— I would say that our surveillance, the checks on radiation and so forth, indicate this."

"It's under control at the moment. But without knowing what equipment has been disabled, how do you know you can keep it under control?"

"Well, I believe we can——" Creitz regroups his forces. "Several things have happened. Number one, we were able to reduce the temperature in the primary system at this time. The radiation levels in the containment vessel have not increased."

"But they're still quite high."

"They are still quite high."

"And the leak is continuing."

"And the radiation level on the outside of the plant," says Creitz emphatically, "is considerably less than it has been."

When we see a scientist at full gallop in his own domain, he often appears larger than life—a specimen of some super-race speaking an advanced language that only his fellow creatures comprehend. In truth, he often knows less about the overall picture of his endeavors than does the average TV watcher. Specialization is the problem. As technology becomes more involved, the technician is forced to know more and more about less and less. Out in the field, the practical problems are so complicated that you seldom run into a general practitioner.

So it turns out that the average roomful of engineers is a collection of educated individuals, each having vast areas of ignorance about what the guy next to him is up to. All this is sand in the gears of heavy technology, and it has been driving Vic Stello nuts since noon yesterday.

For much of the day, Stello's questions about the reactor had to be relayed from NRC headquarters by way of the regional office, into the Unit One control room, then by intercom—or runner—to Unit Two, and back again. At the end of the line, he found himself talking to people who had to be educated for every question. The NRC's man on the phone in Unit One was Chick Gallina, who, by the luck of the draw, turns out to be a health-physics specialist. Gallina can tell you anything you need to know about radiation protection, but he doesn't know a thing about the reactor at Three Mile Island.

So, the information reaching headquarters has been constantly garbled by the translation from one specialist's lingo into another, and all this has combined to make Vic Stello nervous as hell in the middle of the night. A lot of people went to bed exhausted but happy a few hours ago; Stello wasn't one of them. As the gray dawn rises over the NRC complex in Bethesda, he is still stalking the corridors of the Emergency Response Center in the East-West Towers, trying to sort fact from fiction.

Two hundred miles to the north, in the ornate splendor of the governor's office, beneath the icy gaze of William Penn, the state's top officials are discussing pregnancy. The human fetus, with its cells constantly dividing, is extremely susceptible to radiation. Although most of the people in this room are of the male persuasion, they aren't oblivious of this peculiar danger to unborn children. Bill Scranton's own wife, Coral, is pregnant.

Some people in the department of health believe that pregnant women and small children should get out. But the state's own radiation people are saying that the danger seems to have passed, that the situation is under control. Finally, Thornburgh and Scranton decide somebody's got to go down to the island and get the lay of the land; there are too many conflicting reports. Scranton volunteers. He's no nuclear engineer, but at least he can see whether the people down there are running around like chickens with their heads cut off.

The lieutenant governor calls Met Ed and finally gets through to the president, Walter Creitz. Scranton says, "I'd like to come down."

Creitz says, "Great, Senator Hart is coming up from Washington today——"

"No," says Scranton, "this is not a media hit. I really want to come down and get on the island."

"Well . . . I don't know."

Scranton persists. Creitz tries to stall. "I can't be there."

"Fine," says the lieutenant governor, "I'll just go down there quietly."

A short time later, Scranton slips out of the capitol building with his aide, Mark Knause. Knause, a state trooper, takes the wheel of the lieutenant governor's black Buick, and quickly they are speeding south down Front Street, through Steelton and Highspire, along the Susquehanna toward the reactor.

When they reach the plant's north gate they are picked up by an escort. Knause is not allowed to drive the lieutenant governor's car beyond this point, since anything on the island might become contaminated. They switch to a company truck and cross the bridge a little after noon.

Scranton is a powerful young man who comes from a tradition of money and influence, but he feels like a dwarf in this setting. There is nothing here on a human scale. The great cooling towers rise out of the river mists like a concrete overcast, and everywhere around him are the sheer steel cliffs of the windowless buildings.

From the guardhouse, where Scranton and his assistant are fitted out with plastic booties, they are escorted across the yard into the plant and through the maze of passageways that lead to the control room in Unit Two.

It's impossible to enter the control room at Three Mile Island and not be awed. It is enormous. The main panels run for a hundred feet in a giant arc and everything is covered with winking lights. But Scranton finds it a little disturbing. It looks just like the ads for *The China Syndrome*.

On the other hand, he is impressed that there seems to be no hint of panic. There are a dozen people in the room and they all seem purposeful, intent: nobody is shouting, nobody is lunging at the panels. In fact, nobody seems to be touching anything at all. They seem to be just looking at the instruments, checking one thing and another, as if they were probing. A little like blind men inspecting an elephant.

"I'd like to see the auxiliary building," says Scranton.

The people with him are taken aback. Not even Knause was expecting this. Everybody tries to talk him out of it. The auxiliary building has taken on an ominous quality, even for the men who work here. Two days ago it was a benign factory building filled with pumps and tanks. Friendly, almost. Some of the guys used to go down to level 281 on their lunch hour and smoke reefer and play handball against the concrete walls of the tank vaults. Suddenly, it has become a dark cave where humans travel at their peril. The radiation field is 3,000 millirem per hour. This is a dose that would in sixty minutes exceed the three-month limit for a nuclear worker.

"You can't go in there without protective clothing."

"Fine," says the lieutenant governor.

Knause decides to sit this one out. The gov can handle the auxiliary building on his own.

To keep radioactive dust off his skin, Scranton dons several layers of protective clothing that cover his entire body. The suiting-up process takes about twenty minutes. Finally, they fit him with a Scott Air Pack and show him how to tell when he's running out of air. With a health-physics technician leading the way, he trudges off down the hall like an alien invader.

At the access hatch to the auxiliary building, Scranton's companion runs his monitor around the door to make sure nothing catastrophic has happened since the last entry. Then he cracks it slightly and sticks the probe of his Teletector inside. After a moment he nods to Scranton and the two men step in and seal the door.

There is an access hole in the floor of level 305 with a railing around it, and from here the lieutenant governor is able to look down and see the problem three stories below. Although the water down there is deadly, the view from here somehow makes the problem a little more tangible. It looks like a flooded basement.

There is one man who might be able to shed some light on the strange reports coming out of Three Mile Island, but he has been ordered not to get involved. Dr. Roger Mattson is head of the NRC Division of Systems Safety. His group drew up the regulations for the emergency systems. Every major nuclear-safety issue ultimately comes to his personal attention, and he is the government's leading expert on emergency core cooling.

Since the minute he heard about the accident, Mattson has been desperate to get involved. It sounded ominous. But his boss told him not to waste his time; plenty of other people were already on the case. Besides, Mattson and his staff had their hands full with another matter.

It seems that Stone & Webster Engineering Corporation, one of the major nuclear-plant architects, somehow used the wrong computer program to determine whether their plants could withstand earthquakes. There are five nuclear plants —all on the East Coast—that are affected by the mistake, and now the whole mathematical basis for the design of these structures will have to be reexamined. In the meantime, the NRC more or less has to order the five plants closed until Stone & Webster can make new calculations.

It has created a furor up on Capitol Hill. Some congressmen are saying this incident proves the nuclear people don't know what the hell they're doing. Others are saying that it's insane to shut down the plants, since the odds against an earthquake are astronomical to begin with. Mattson is digging through the details of this issue, but his mind is elsewhere. For a day and a half now, his anxiety about the situation in Pennsylvania has been building. Finally he turns to one of his aides. "The hell with it," he says. "If they won't invite me over, I'll go on my own."

As an excuse, he bundles up a bunch of papers on the five-plant shutdown and heads for the Incident Response Center, four blocks away in the East-West Towers. When he gets there, he finds Vic Stello still in charge. From the look of his clothes, Stello has obviously been here through the night.

Mattson and Stello have been working together since the days of the old Atomic Energy Commission and they have great respect for each other, but their personalities could not be more opposite. Mattson is open, outgoing, soft-spoken. Stello is tough as nails. Mattson—bearded, dark-haired, thoughtful—looks like a scientist; Stello looks like a line-backer. And sometimes he behaves like a line-backer. If you don't have your facts straight, he runs over you. But right now, he's very glad to see his old pal.

"Great!" says Stello. "I need you. Here's what I want you to do." He tells Mattson about the unusual thermocouple readings they are getting from the core. Some are so high —in the range of 2,300 degrees—that nobody believes they're real. But Stello is having second thoughts. "Don't talk to me about 'why,' " he tells Mattson; "just go off and believe the thermocouples and tell me what they mean."

Mattson goes out to find a quiet place where he can dig into this. But out in the hall, he runs into Denny Ross, one of Stello's assistants. Ross has a sheaf of documents in his hand. "I've gone back and read some of the reports from the other partial core melts," he says, "where people lost fuel assemblies in test reactors, and at Fermi One. I wanted to see if there was anything I could learn." Ross holds up the documents.

"There's a common thread that runs through all of this. If the operators had believed their instruments, they would have made better choices. For what it's worth, history teaches us: believe your instruments."

It is a lesson any airline pilot knows by heart. "Believe your instruments." When you begin to argue with the com-pass, it is the first sign of panic.

But Joe Hendrie is an engineer, not a pilot. He's a good engineer, but he's never had the experience of tumbling through the air with only a set of gauges between life and death. His instincts tell him, as they have told almost everyone else in the NRC, that this 2,300-degree thermocouple reading is impossible.

Joe Hendrie is the top man at the NRC. As chairman of the five-man commission, he is the leader of the band, so today he is heading the contingent of experts who have been called on the carpet by Congress.

In the Rayburn Building, the paneled hearing room is packed. The back wall is lined with TV cameras. Normally, Mo Udall's Subcommittee on Energy and the Environment isn't such a hot ticket, but this morning, everybody wants to know what happened up in Pennsylvania. Chairman Hendrie, surrounded by staff aides, sits at the center table, hunched over the microphone, trying to explain. What he has to say is reassuring.

"At this point, we think the plant is in a very stable configuration and we think, from a system standpoint, we understand very well what is going on."

Congressman Symms of Idaho wants to believe what Hendrie is telling him. He's no critic of nuclear power. He believes it holds the key to energy independence. But he would like to have his doubts eased about all these reports of radiation. Symms says, "You mentioned there were 3,000 rems at the dome . . . That would be at the top of the containment building?"

Hendrie says, "That is inside, behind four- or five-foot concrete walls."

"That is not outside?"

"No," says Hendrie. "Good heavens, no."

"Is that the highest reading, at the dome?"

"There is an instrument at the dome, Mr. Symms, which I understand has reported a series of readings of some thousands of R per hour dose radiation level. But these levels seem to be inconsistent with the reading of other instruments . . . I am inclined to think that we have got a faulty readout in that instrument."

Unfortunately, there are no airline pilots in the room.

But there are skeptics. Congressman Jim Weaver, a Democrat from the Oregon coast, has spent a considerable amount of time on Capitol Hill. In his day, he has seen more than one federal bureaucrat scratching dust. He doesn't like the way this story fits together.

"Mr. Hendrie . . . we have heard that the primary system failed, the secondary system failed, the valves failed, and there was possibly human error. You say that this was the most serious core damage that we have ever had. How close did we come to a meltdown, to a Chinese syndrome?"

"Nowhere near."

"Could it have happened?"

"In principle, Mr. Weaver, we could have a meltdown situation at any operating plant going on right at this moment. But the chances of that happening are really very small, diminishingly small."

Weaver pales. "Mr. Hendrie . . . you and other members today have constantly said 'apparently . . . ,' 'we are not sure what is going on . . .' 'we think there is enough water to keep the situation going along . . .' 'apparently an abnormality occurred . . .' " He leans to the microphone. He is grave and shaken. "You do not really know what is happening, do you?"

While Chairman Joe Hendrie is briefing Congress up on the Hill, the four other commissioners are minding the store back at the executive office near the White House. In his corner suite, Commissioner Peter Bradford paces the floor reading over the latest NRC official update. From the way this thing is written, it's clear that it's all over but the shouting.

. . . As of 3:30 P.M., the plant is being slowly cooled down with RCS pressure at 450 psi, using normal letdown and makeup flow paths. The bubble has been collapsed in the A RC loop hot leg, and some natural circulation has been established. Pressurizer level has been decreased to the high range of visible indication, and some heaters are in operation . . .

Media interest is continuing . . .

These people have a talent for understatement that has evolved into an art form.

Just then Bradford's technical assistant, Hugh Thompson, walks in with some papers in his hand. "I just found this in my in-box," says Thompson. He rolls his eyes. "This is gonna be a memo that somebody wishes he'd never written."

It is a letter of transmittal from the NRC technical staff about a problem that has been stirred up by some inspector out in the Chicago regional office. It seems this fellow had some concerns about the safety of the Babcock & Wilcox reactor. But the staff looked into it and told him his fears were groundless. The only reason they are passing on the report is that the inspector demanded it.

Bradford grabs the memo.

Creswell!

He sinks into his chair. Five days ago, Jim Creswell sat right there, in this office . . .

Hugh Thompson reaches for the phone and dials the staff headquarters in Bethesda. He gets through to Dudley Thompson (no relation), one of the staff executive officers. Apparently the temptation to continue burying the bodies is still in full sway. Dudley Thompson confesses, "It's a trifle embarrassing to I&E headquarters* that is, that the opinions expressed in that evaluation memorandum as it—— We continue to believe that the inspector is placing undue emphasis on a subject that is being addressed in other areas adequately, and we sort of have egg on our face if there is any direct correlation between his observations, our nega-

*Inspection and Enforcement Division of the NRC

tive reaction to them, and the Three Mile Island situation."

"Yes."

"As a matter of fact, it crossed my mind very briefly that I ought to give Moseley a chance to recall his memorandum, since it's only dated yesterday, and I have not taken any action on it."

But the commissioner's aide warns him against it. "Apparently this one is out public, as far as I know."

The staff man in Bethesda is not sure which way to turn. "I would hate very much for us to have an evaluation that says, 'Gee whiz, now that we look at it, you're absolutely right, and this is a very serious problem,' dated the day after or two days after this thing took place at Three Mile. That's just blatantly obvious."

The staff executive agrees to check into the situation and get back to the commissioner right away. After a frantic round of interoffice hot-potato out in Bethesda, Dudley Thompson calls back with the word.

"Hi. We've talked the whole thing over, and as much as we'd like to have it otherwise, we're going to take the position. You'll have it all . . . including the original incoming on which we acted a couple of months ago . . . in the negative sense."

"Uh-huh."

"There appears to be some superficial similarity between the concerns identified by Creswell and what has transpired, but it's not at all clear that it's a one-for-one correspondence . . . I just didn't want the commissioner to look at this package and conclude that reviewers of a position that is superficially very appealing in light of today's events were a bunch of idiots who were not listening to a supersmart inspector, because that's not the case."

Commissioner Bradford, however, is a lawyer, and he has no difficulty understanding the ominous implications for the system. He looks up at his aide. "Here was a guy trying to tell us a year and a half before it happened that this kind of incident could take place, and nobody listened . . ."

It is boron that keeps the reactor from going critical. Boron in the cooling water absorbs loose neutrons. If there is not enough boron, the chain reaction could restart, and if they didn't catch it quickly, there would be little to do but run for it. Normally, the operators are required by law to sample the main coolant every few hours. But since dawn yesterday that's been out of the question. The last reliable sample is over twenty-four hours old.

Nobody at the plant—where they are surrounded by radiation—is particularly eager to get a new sample, but at the corporate headquarters in far-away New Jersey, some of the top technical people are getting eager to find out what that water looks like. And so the word filters down through Jack Herbein across the river, through Gary Miller in the shift supervisor's office, through radiation supervisor Dick Dubiel, to foreman Pete Velez: "Get the sample."

It is inevitably a guy like Velez who must pick someone to go below decks and bring out the unexploded bomb. And since Velez is an old navy man who would rather croak than look like a chickenshit, he inevitably picks himself.

Ed Hauser will go with him. Hauser is the chemistry foreman, and he is the man who will actually have to take the sample while Velez monitors the radiation. Hauser is a quiet young man with a wife and two kids and a new house in a rolling meadow not fifteen minutes from here. Like Velez, he never imagined that it would one day come to this. Everybody always said the safety systems would prevent it.

The two men carefully check their equipment and then inspect each other. Suiting up takes nearly half an hour: coveralls, wet suit, four layers of gloves—everything taped together: boots to coveralls, gloves to wet suit. Finally, they put on face masks, adjust the breathing regulators, and are off. Velez takes the lead with his Teletector extended the full twelve feet so that they won't stumble onto any surprises.

The health-physics lab next to the hot-machine shop was abandoned in the early hours of the accident, and when Velez opens the door, Hauser peeks in on a sci-fi scene right out of the Bermuda Triangle. Hats on the hat rack, jackets hanging, phones off the hook, a pot of coffee on the table— everything in place but the people. At least there aren't any burning cigarettes in the ashtray.

Like Velez, Hauser respects radiation, and he's always been careful to hold his exposure to a minimum—at least until yesterday. Then he picked up 600 millirem in one shot while trying to run a check on one of the boilers.

Behind that door is a hooded sink with about twenty-five valves; they all have to be properly lined up to get the water flowing. To share the risk, Velez volunteers to split this task with Hauser, but as they talk it over, Hauser realizes that Velez doesn't know the system. He'll have to do the valve lineup by himself.

After Velez makes a preliminary survey of the room, Hauser pauses at the door, memorizing his act. Velez checks his watch and shouts, "Go!" Hauser dashes in, straight for the back panel behind the sink. With both hands moving like fury, he turns some fifteen valves; a quick glance at the pressure gauge shows that the water is moving. He heads for the door.

The sample pipe that runs from here to the reactor and back is small, naturally—finger sized—and it's a long run. So it takes about forty minutes for a fresh sample to reach the lab. In the meantime, they wait in the corridor outside. And as they wait, the radiation builds. Finally Velez goes in to check the level at the sink. From twelve feet away, he touches the tip of the probe to the sample line. The needle jumps to the pin. Quickly, he backs out and slams the door.

"It's over 1,000 rem on contact!"

Hauser grits his teeth. This wasn't part of the deal. But as he gets ready to go in, Velez volunteers to get the sample; he doesn't want Hauser to take more than his share of the dose. Hauser is relieved. He explains how to draw off the pressure and which valves to open. He checks his watch and gives Velez the signal. Velez rushes in—and he's back out in a flash.

"It looks like Alka-Seltzer," he says.

"What?"

"I crack the valve and out comes this milky stuff— fizzing." Velez says he's never seen anything like it. He figures Hauser screwed up the valves.

Cautiously, Hauser slips in to the sample room. His eyes race over the valve lineup. Everything is right where it should be . . . No sense wasting time. He approaches the sink, slips a glass beaker under the hose, and opens the valve. Instead of clear water, he sees a frothing yellow stream. He realizes it's from all the chemicals that were injected by the emergency systems.

But there's no doubt this is core coolant. The radiation level in this fizzing glass is 1,250 rem—a level so high you can actually feel the tingle of the invisible rays by touching the beaker. If he simply held it in his hands, he would be a walking dead man in twenty-four minutes. Hauser shuts off the sample line, picks up the beaker, and in gingerly fashion carries it to an adjacent room, where he sets it behind a lead shield.

Now for the tricky part. What he's after is a boron count. He wants to know the percentage of boron in the water, and this is a little like checking the chlorine in a swimming pool. First you have to add acid to neutralize the sample or you'll get a false reading. But what everybody seems to have forgotten is the reason all these chemicals were injected into the coolant water in the first place: to trap iodine. As Hauser adds acid to neutralize the sample, the deadly gas is released, and he is enveloped by an invisible cloud of radioactive iodine. Also, he's been in here longer than he thinks. In the confusion, Velez forgot to time his last entry.

Hauser finishes up as quickly as he can. As soon as he gets out into the corridor, he grabs for his dosimeter. But because of four layers of gloves, the pen-sized instrument slips from his fingers. He stoops to pick it up and holds it to the light. The needle reads an unbelievable 5 rem. "Must have broken when I dropped it," he says.

The boron count is okay, and as soon as they phone the numbers up to the control room, both men strip, put on new coveralls, and head for the shower in the security building. Hauser is exhausted when he sinks to the bench in the locker room, but as he starts to take off his boots, he hears a rattling noise at the end of the bench. Somebody forgot his portable radiation monitor. Hauser gets up to turn it off. But as he reaches for the instrument, it squeals in alarm.

Wide-eyed, he backs away. Omigod.

One of the technicians goes to him and tries to run a hand frisker over his body, but the needle pins at the top of the scale. "Get in the shower!"

For the next hour, Hauser drenches and scrubs himself with industrial caustics and solvents. When he gets out, he looks like a prune, but when they frisk him he's still too hot to read. Back in the shower, as he rubs his body raw over the next six hours with Radiac Wash, alcohol, soap, and acetone, he catches himself wondering from time to time if this is really worth it.

For six hours, the water trickles down the shower drain, carrying with it what little contamination Ed Hauser is able to scrape off. And now this waste water is itself contaminated. Normally, the water from the showers, the toilets, and the floor drains is only mildly radioactive. The contamination is within the NRC limits, and the water is simply treated and dumped into the Susquehanna. But when Gary Miller declared a site emergency yesterday, the river drains were closed as a precaution. Now there are 400,000 gallons of waste water down in the basement of the turbine building, and they've got to get rid of it before it starts backing up through the floor drains.

A couple of technicians are dispatched to sample the tanks, and they report back that the reading is within limits. There are several NRC inspectors present in the control room, and after a quick discussion, they give permission to release. A call is made to the state to notify the Department of Environmental Resources, and at 2:30 P.M. the river drains are opened.

Unfortunately, this whole transaction took place on the local level in a fairly routine fashion. The various helmsmen of these vast state and federal institutions were not informed. And when NRC chairman Joe Hendrie finds out from a newspaperman that Three Mile Island is releasing radioactive water into the Susquehanna, he nearly chokes.

He fires a call through to staff headquarters and gets Edson Case, one of the executive officers.

"Ed?"

"Yeah."

"What's going on with this dump down at Three Mile into the Susquehanna?"

Case is caught flat-footed. "Mr. Chairman, we aren't certain of that. We had a report at about 2:30 this afternoon that they were releasing some . . . controlled release into the river . . ."

"I thought they weren't going to do things like that without letting us know."

"Well," stammers Case, "they let—as I understand it, they let us know that they were dumping the—— They maintained, I gather, that it was in the licensed limits."

Hendrie doesn't give a damn about the details. "If Three Mile were operating normally then . . . that would be all right. In the circumstances, why, the impression everybody will have is that he's dumping the contaminated water into the river."

"Bad PR, agreed."

Hendrie is livid. "Supposedly we've got a team down there that's keeping track of things and I'm going around telling congressmen we have good, close communications and that we and the state people and everybody else are working closely together so that we all know what's going on, all agree on the steps, and I don't find it that compatible with him just deciding, 'What the hell, I'll dump 100,000 gallons . . .' "

While the chairman blisters his hide, Edson Case frantically signals other staff people to get on the problem. A few moments later, he's able to tell Hendrie, "The word has now gone—gone back through our chain to the licensee to stop."

What about the state people? asks the chairman.

"The state was, as I understand, was aware—aware of it."

"Like us? They were just told he was dumping it?"

"Yes."

"Jesus Christ."

The word tumbles downhill through the organization. When it reaches the island, the control-room operators look at each other, shake their heads, and shrug. As ordered, they close the river drains, but they know it won't be for long. The tanks down in the turbine building are good for only a few more hours. Then the water will overflow the sump and trickle out across the road and run into the river anyway.

Very shortly, Edson Case and the NRC staff in Bethesda

realize that the release is not only inevitable, but that the NRC had authorized it. This puts the agency in a rather awkward position. They have already put out a press bulletin saying the release was unauthorized and that Metropolitan Edison had been ordered to stop immediately.

Somehow—and soon—somebody's going to have to bite the bullet and say the release is on again. As a face-saving maneuver, one of the NRC staff people comes up with a solution. Why not get the state of Pennsylvania to authorize the release? Let the state say it's okay and the NRC can stay in the background. Everybody in Bethesda likes this idea. Now the trick is to sell it to Pennsylvania.

The handling of this ticklish problem falls to Karl Abraham, the Region One public-affairs director who just happens to be in the state-capitol building in Harrisburg. Since this morning, Abraham has been the NRC liaison man at the state house, and over the last few hours he has developed a rapport with the governor's press aide, Paul Critchlow.

After a frantic round of conference calls with Bethesda, Abraham goes into Critchlow's immense office just outside the governor's suite and broaches the problem. He tells Critchlow that the NRC has no objection to releasing the waste water in the turbine building, but it's really up to Governor Thornburgh. So if it's okay with the governor . . .

But Critchlow used to be a big-city newspaperman, and Thornburgh, after all, didn't hire him because he was stupid. "Karl," he says, "I'm not gonna put the governor in a press release that talks about 400,000 gallons of contaminated water going down the river and say it was his decision when it's not. It's very clear to me it's your decision."

Abraham leans across the desk. "How about the lieutenant governor?"

There is a natural bias to the human thought process, a subtle set to the mind that affects us so gently we seldom realize how it might alter our judgment. The truth is, we tend to remember the good parts and leave out the bad. This automatic self-protection filter keeps us from being paralyzed by total recall of all the stupid things we've ever done. But in a crisis, it tends to constrict the flow of information.

The bad news—the hydrogen explosion in the containment, the 2,300-degree thermocouple readings, the staggering 40,000-rem indication from the containment dome—all this has been dismissed or dropped from the discussion as it moved upward in the Met Ed organization. At the company offices in Reading, nobody has mentioned any of this in some time. And at the New Jersey corporate headquarters of the parent company, the subject has hardly come up.

So when Dick Wilson, the top troubleshooter from corporate headquarters, arrives at the island, he expects to find a slightly messy plant on its way to cold shutdown. He's been sent here to find out what went wrong and how quickly they can get the plant cleaned up and back on line. He is dismayed when the guard hands him a Scott Air Pack and tells him breathing apparatus is required on the island.

Wilson gets another jolt when he reaches the control room. He had hoped to start interviewing Zewe and Frederick and the other operators who were on shift when the accident began. But Gary Miller tells him it's out of the question; his people have their hands full.

Wilson checks around the plant with growing apprehension as he talks with one group and another. When he hears about the 1,250-rem coolant sample, it begins to dawn on him that he has been sent here on the wrong mission. Not only is the reactor not being shut down; there is no one here with a clear idea of how to go about it. He finds a phone and gets through to his boss, Bob Arnold, in New Jersey.

"The situation is much different than we thought," says

Wilson. "The scope of the accident is much greater." All in all, it's a grim report. The damage to the reactor is substantially more than anyone realized. There is a serious radiation problem on-site. George Kunder has requested engineering assistance in the control room. It's clear to Wilson that these people need all the help they can get. He recommends that the company start trying to locate all the technical and physical assistance they can lay their hands on. Finally, he says to Arnold, "You should plan on being at the site as soon as you can."

Unfortunately Chick Gallina has quite the opposite impression. Gallina is an NRC investigator on duty in the control room, and from his standpoint things are looking up. It is his understanding that all this radiation has been coming from the water that overflowed into the auxiliary building in the early hours of the accident. By now most of that water has been pumped into storage tanks. What little is left on the bottom floor has been covered with plastic sheets. To Gallina, this means the worst is over.

So, this is the official impression he carries off to the state-capitol building in Harrisburg late Thursday afternoon. Gallina is the man the NRC has picked to brief Governor Thornburgh and his aides.

Based on Gallina's optimistic update and Scranton's first-hand reassurance that the control room was "calm," Thornburgh's press aide bangs out a statement, and the governor faces the cameras in time for the late news.

"I would like to direct my initial remarks to the people of central Pennsylvania.

"I believe, at this point, that there is no cause for alarm, nor any reason to disrupt your daily routine, nor any reason to feel that the public health has been affected by events on Three Mile Island . . .

"I have spent virtually all of the past thirty-six hours trying to separate fact from fiction about this situation. I feel that I have succeeded on the more important questions . . ."

Thornburgh is flanked by the lieutenant governor, several state experts, and Gallina, from the NRC. During the question-and-answer session that follows, everybody is cautiously positive. But after listening to all this mutual reassurance, Gallina gets a little carried away. He ends the press conference by saying flatly, "The danger is over."

Scranton's stomach turns over. He glances at the governor and can see that Thornburgh has had the same reaction. There is no way at this point that anyone can say with that kind of assurance, "The danger is over." What had been, moments ago, a slowly building tide of confidence in the NRC is now ebbing. These bastards are as bad as Met Ed.

Tonight the Pennsylvania state house looks like a casino —lights in every window, cars arriving, people running up the white marble steps. It is late, long after the last Big Mac and take-out pizza has been pitched into the wastebasket, but Paul Critchlow is still at his desk just outside the governor's office.

He punches the button for the next call. "Critchlow."

The quiet voice at the other end has an edge that brings Critchlow upright in his chair. It is an NRC inspector he met the day before. The man is calling from his motel room. He asks Critchlow not to use his name.

"I feel the governor and the public have been given a slightly misleading impression about how serious it is," he says. "Inspectors are now coming off shift from the plant and they are telling me the core damage is indeed very bad."

"What does this mean?"

"It means the long-term consequences might be extremely long-term."

"How long?"

"Maybe years."

Critchlow is stunned. This is the first hint of the actual dimensions of the problem. "What are the short-term consequences?"

"They could also be very severe," says the voice.

In a former life, Critchlow was a forward artillery observer in Vietnam, so he recognizes this feeling in the pit of his stomach. On his way to let Thornburgh in on the bad news, he stops for a second to tell his assistant, Pat. He is concerned about her, and who could blame him? She is a strikingly beautiful young woman whose dark eyes would stop almost anybody in his tracks. She was a campaign volunteer when Paul met her, and they've been together ever since. He wants to know if she would consider getting the hell out of here. "You ought to think about it."

"No," she says, "I don't want to leave unless you do."

Shit. Just like in the movies, they are trapped. Critchlow can't leave until the governor leaves, and the goddamn governor will be the last man out.

". . . Yea, though I walk through the valley of the shadow of death, I will fear no evil, for Thou art with me . . ."

Mrs. Ed Hauser is a Christian; she believes that God has a plan for us all. Surely there must be some reason why Ed is coming to bed with gloves on his hands and a plastic cap taped to his head.

14

Friday, March 30

Recalling an island fortress under siege, the glaring emergency lights reveal a constant flow of fresh troops moving across the north bridge. In addition to the double shifts Met Ed and the NRC are laying on, people are coming from half a dozen federal agencies; there are engineering consultants, radiation specialists, insurance men, and a growing number of private contractors. As the newcomers arrive at the security building to pick up their radiation gear, nobody can resist a glance at the awesome dome on the south end of the island.

Bathed in floodlights, its gray walls now streaked with rain, the mighty containment building looks used, like a leaking dam. But it is the one safety system that actually did its job. If the containment building had not held on Wednesday afternoon, when the hydrogen explosion shot the pressure up to twenty-eight pounds per square inch, the disaster would have been instant. If there had been one faulty weld in the hundreds of critical welds, one weak pocket in the acres of concrete, Gary Miller and Jack Herbein might never have made it to the governor's office. They could have died on the spot along with their wives and children and thousands of others.

The containment did hold, however, just as they said it

would. If there is one thing the engineering profession has figured out over the last several centuries, it is how to build a wall. But as Gary Miller discovered in the opening hours of the incident, there are other, less spectacular ways for radiation to find its way out of the reactor. By overlooking a single set of switches, they accidentally pumped several hundred thousand gallons of reactor coolant out of the containment into the basement of the auxiliary building. Just as the operators get that situation under control, they find an even more ominous hole in the dike.

In normal operation, a small amount of water is continually drained out of the reactor into a purification-and-treatment system, where it is stored in holding tanks and pumped back into the reactor as needed. All of the pumps and tanks for this purification system are located outside the containment in the auxiliary building.

Now, the reactor coolant is loaded with radioactive gas. As Pete Velez discovered yesterday, it looks like seltzer water. As the water comes out of the high-pressure reactor system into the low-pressure purification system, the gas escapes from the water. It is collecting in the storage tank inside the auxiliary building.

Naturally, there is a limit to the amount of pressure this tank will hold. The limit happens to be eighty pounds per square inch; at that point a safety valve will open and dump the whole thing, water and all, out of the system into a waste tank. They can't afford this, since they are running short of cooling water as it is.

So, for the past twenty-four hours, they have been "burping" the gas out of this tank into another container, where the gas is compressed and stored until the radioactivity decays. Unfortunately, the pipe connecting these two tanks has a leak in it. Every time they "burp" the system, the intensely radioactive gas leaks into the auxiliary building and out the stack.

Jim Floyd was supposed to take over the control room at midnight, but he actually got here an hour early. The situation is now so involved that it takes about an hour of overlap for the oncoming shift to pick up the reins. The outgoing crew explained to Floyd the method they were using to vent the storage tank; it's a tricky process. Every time they hit the valve to blow out the gas, the radiation level in the auxiliary building takes a sudden jump. So just before each "burp," they have one of the men suit up and run into the auxiliary building to start the gas compressors. In this way they've been able to suck the radiation through the vent pipe as fast as possible and at least keep the leakage to a minimum.

But in the small hours of the morning, Floyd can see they are losing ground. The gas is building up faster than they can handle it. There is less and less time for the business of getting a man in there to start the compressors. Then Floyd notices that the water level in the main waste tank is slowly increasing. This tank is almost ready to overflow as it is, but Floyd feels a more pressing anxiety.

Where the hell is this water coming from?

Quickly he traces the flow and realizes that the safety valve in the storage tank must have opened. There's only one way to stop it now. He's got to vent the gas pressure off of that tank or lose his reserve cooling water.

It's a grim choice, but it takes only a second for Floyd to make up his mind. If he vents the tank, some people might get hurt. If he doesn't vent the tank, he could be setting the stage for a meltdown.

"Open the vent!" This time, Floyd orders them to keep it open. His next call is to the emergency command center in Unit One. "I need a helicopter over the plant stack right away. We've got a release."

As the chopper rises over the island to monitor the escaping radiation, a disturbing thought flutters through Jim Floyd's mind. They've been using that vent valve like a swinging door for the past twenty-four hours. What if it sticks open?

Floyd is one of the few people in this plant with any gut-level sense of the real damage downstairs. He knew from that first phone call down in Lynchburg that the core was fractured. By now he's seen enough to realize that the damage is far worse than anybody is letting on.

Floyd still believes they'll be able to work their way out of this mess. But he keeps thinking about his son, a volunteer fireman who is sleeping tonight at the Middletown fire station, not 3,000 yards from here. It's one thing for him to stand here bravely on the bridge, prepared to go down with his ship. But there are people "out there"—wives and children.

He calls the Pennsylvania Emergency Management Agency to make sure they are ready for the worst.

Colonel Oran Henderson, the head man at the state agency, is standing near the phone when the call comes through. One of his aides picks it up. The man listens with growing concern, then turns to Henderson. "This guy is going ape. He says that they have a reported reading of 1,200 millirems at 600 feet. And they're recommending that we get prepared for evacuation downwind. They are prepared to evacuate noncritical personnel from the island now . . ."

Henderson says, "Call the weather bureau and the airport. Find out which way the wind is blowing." Then he shouts to his secretary, "Get me the lieutenant governor."

Although it seems like a military bunker, the NRC's Incident Response Center in Bethesda is not buried underground. It is in the middle of an office building that the agency shares with several businesses. But it has no windows; just blackboards, wall charts, and giant tape recorders slowly turning amid long rows of telephones. The only way the people jammed into this pressure cooker can tell it's daylight outside is by checking the twenty-four-hour clock.

This was supposed to be the command post for the crisis managers, but it isn't working. Like the French generals of the Maginot line, it seems, the NRC has planned for the previous war. The agency's last close brush with disaster was the Brown's Ferry fire—a thriller, for sure, but the whole thing was over in a matter of hours. Nobody ever imagined an accident that would just keep running on and on and on. So, the formal procedures have been abandoned, and the executive session has disintegrated into a cluster of specialized discussions. The command post looks more like a newspaper city room with gaggles of rewrite men huddled around half a dozen different stories.

Some of these people, shirt-sleeved, unshaven, haven't been home in three days. Lake Barrett is with the environmental branch of the agency. He spent most of the night trying to estimate the overall population dose since the accident began. In the process, he has become concerned about all that radioactive gas that's still coming out of the system. For the moment, the operators are able to compress most of the gas into waste tanks, but what happens when the waste tanks fill up?

Just then, somebody hands him a telex from Region One. It's a report from an NRC inspector on-site:

> . . . excessive gas pressures in the makeup tank, which was directed to the waste-gas decay tanks, which were full. The waste-gas tanks were being released to the stack. Pennsylvania Civil Defense was being notified by Licensee . . .

In fact, this Telex is in error. The tanks are not yet full. But the possibility ignites Barrett's imagination. In his head, he does a quick calculation and is staggered by the numbers. Based on his data from yesterday and his estimate of what kind of incredible crap must be in those tanks, Barrett figures that a person at the plant fence could expect a dose of 1,200 millirem an hour.

At that moment, one of the top agency officials is working his way through the room and overhears Barrett's agitation. In less than a minute, the environmental expert finds himself thrust into the glass-enclosed executive office. The man ushering him in shouts for attention, and in the sudden quiet, every big shot on the NRC staff is looking at Lake Barrett.

Barrett is a scientist, not a public speaker, but he's too tired to be intimidated. He blurts it out. "I believe under the current conditions the off-site reading might be 1,200 millirems."

"Oh my God!"

"That's over the EPA protective-action guidelines."

They are shocked. Although it's finally clear to these men that the plant is badly damaged, the off-site radiation so far has been fairly innocuous. This is something else. This is where you start moving people.

And at this moment, a bombshell is coming in from Harrisburg. Lieutenant Governor Scranton has heard from Colonel Henderson about the possibility of immediate evacuation and something about a 1,200-millirem release. Scranton was at home when the call came, so he checked with his office; they knew nothing about it. The NRC's man in Harrisburg, Karl Abraham, was told to call headquarters and find out what the hell is going on.

Since Karl Abraham is a PR man, he naturally calls the PR man at headquarters, Joe Fouchard, who is standing in the executive conference room across from Lake Barrett. Fouchard is stunned by what Abraham tells him. He shouts for attention. "The licensee is now measuring 1,200 MR per hour."

None of these people realizes that this 1,200-millirem

reading—the reading that got here from the control room by way of the state house—is a measurement above the *plant stack*, not at the plant fence. But the way this number matches the prediction they've just heard gives everyone the jitters. Suddenly, nobody gives a damn about the details.

"It is time to bite the bullet."

Harold Denton, the director of nuclear-reactor regulation, just got here from the "Today" show. It was his first television experience. He rather enjoyed it. He has just finished telling the country that everything was okay. Now he's going to have to tell them to run for it.

"If we are going to err, let it be on the side of public safety," says Denton. He turns to Barrett. "How far should people be moved?"

Barrett is paralyzed. A couple of minutes ago he was standing outside talking to himself, and now all of a sudden he's become the agency's expert on evacuation. He doesn't have an answer; there are too many variables: wind speed, population boundaries . . . "I can't recommend the specific distance."

Harold Denton, they say, can be quite ruthless, and right now he's in no mood for mealy-mouthed answers from the junior staff. He levels his gaze. "How far should people be moved?"

It's clear that Barrett at least has to make a stab at it if he's going to leave the room in one piece. "It is hard to tell," he says, "but ten miles is more than enough." As soon as the words are out of his mouth, Barrett realizes that this includes part of Harrisburg.

After a quick argument over the details, the NRC senior staff agrees to recommend immediate evacuation. Harold Denton cuts off the discussion and turns to the liaison officer, Doc Collins. "Call the state of Pennsylvania and recommend the implementation of this precautionary evacuation."

Immediately, everybody breaks for the telephones in

search of commissioners, scientists, and assistants, and Lake Barrett stands alone like an abandoned cigar-store Indian.

As Doc Collins dashes from the room, his first decision is to violate the procedures he himself worked out with the state. According to the book, he's supposed to notify the State Bureau of Radiation Protection; but Collins has been in a running battle with those people, so he decides to skip them for now. Instead, he calls the Pennsylvania Emergency Management Agency and immediately gets through to Colonel Oran Henderson.

"Oran? Have you heard the latest report from Three Mile Island?"

"Yes."

"What do you know?"

"They have just had a 1,200-millirem release."

"That is the same information that we have," says Collins. "Colonel, we are recommending that you execute immediately a ten-mile evacuation around Three Mile Island."

Unlike Chicago's late Mayor Daley, Bill Scranton is not the kind of politician who likes to call attention to himself. He has never used the flashing lights on his state limousine. But this morning, he's willing to make an exception. As he races from his home in Indiantown Gap, some twenty miles out of Harrisburg, he tells his driver, trooper Mark Knause, to hit the siren.

Wailing through the Friday-morning rush hour at high speed, Scranton's black Buick screeches to a halt across from the capitol mall. The lieutenant governor jumps out and sprints for the transportation building, which houses Colonel Henderson's Emergency Management Agency. The sidewalks are full of people buzzing about rumors of evacuation. He's not surprised to see the same scene inside. The lobby is jammed to the walls. But just as Scranton prepares to calm the crowd, he realizes that everybody is waiting in line. Today is the last day for motor-vehicle registration. Three Mile Island be damned; these people want their license plates.

The nation's capital always has some of the characteristics of a swamp—the British embassy lists it as a tropical post—but it's not often they get a heat wave like this in March. The secretaries hurrying to the office buildings along H Street have the silken visibility of summertime; but eleven floors above the sweltering pavement, the atmosphere is wintry.

Here the five presidential appointees who run the Nuclear Regulatory Commission have gathered in Peter Bradford's office to listen on the squawk box to the bits and pieces of bad news coming from staff headquarters in Bethesda.

Hendrie, in his middle fifties, has been in the nuclear business since he got his doctorate from Columbia. He has always worked for the government, first at Brookhaven National Laboratories, then with the old Atomic Energy Commission. He is the only one of the five commissioners with any clear idea of how a reactor works.

Peter Bradford is a lawyer. Richard Kennedy is an economist who got here by way of Harvard Business School, the National War College, and the National Security Council staff under Kissinger and Haig. Both Victor Gilinsky and John Ahearne have engineering-physics degrees from Cornell—they graduated within a year of each other—but their understanding of nuclear-power plants is much more academic than Hendrie's.

These five gentlemen don't necessarily like each other, but everybody respects Joe Hendrie's practical knowledge. So, even though technically he has no more authority than any of the others, they all defer to him. This morning, however, he's having trouble making up his mind, and the voice on the speaker phone isn't much help.

Harold Denton is calling from Bethesda. He seems to be almost as much in the dark as his five bosses. Hendrie wants to know if Denton has heard anything from the chief investigator they sent up to the island yesterday.

"No, I haven't," says Denton.

"Is he on top of it?"

"Well, I sure hope so, but he is not in the dialing-communications line and I have not been able to reach him."

Then another more urgent voice cuts in. It is PR man Joe Fouchard. "Mr. Chairman. I just had a call from my guy in the governor's office, and he says the governor says the information he is getting from the plant is ambiguous, that he needs some recommendation from the NRC."

Denton ignores this, and so does everyone else. "It's really difficult to get the data," he continues. "They opened the valves this morning . . . and were releasing at a six-curie-per-second rate before anyone knew about it. By the time we got fully up to speed, apparently they had stopped . . . We did advise the state police to evacuate out to five miles, but whether that has really gotten pulled off we'll just have to——"

"Well," says Fouchard, "the governor has to authorize that, and he is waiting for a recommendation from us."

For twenty years now, Joe Hendrie has been convincing people that this moment would never come. Now it is here. Like a cancer patient who refuses to hear what the doctor is saying, the chairman's turns his mind to other things. He locks in on the details. He wants to talk about the radioactive plume—its size, its shape, and the possibility that it has already passed and they won't have to do anything about it at all.

But in Bethesda, Denton is not reassuring. "We don't know how long, but if it was a continuous release over a period of an hour and a half which . . . is kind of a lot of puff," he says. "I think the important thing for evacuation is to get ahead of the plume—is to get a start rather than sitting here waiting to decide. Even if we can't minimize the individual dose, there might still be a chance to limit the population dose."

To the commissioners, this all seems quite bizzare and unbelievable. Yesterday, the accident was over. They were sufficiently unconcerned about the situation in Pennsylvania that they spent the whole afternoon on an entirely unrelated matter. This morning, all of a sudden, here are people talking about . . . evacuation?

"Mr. Chairman," says Fouchard, "I think you should call Governor Thornburgh and tell him what we know."

Another commissioner breaks in. "Well, one thing we have to do is get better data. Get a link established with that helicopter to make sure that from now on we get reasonable data quickly."

Fouchard tries to get everybody to stick to the point. He has a picture in his mind of the governor of Pennsylvania, and it's not pleasant. "Don't you think as a precautionary measure there should be some evacuation?"

"Probably," says the chairman, "but I must say, it is operating totally in the blind . . ."

Hendrie is at sea. He doesn't know what caused the release, how big it is, or what the situation is now. And from Bethesda, Denton admits that he can't give them any assurance it won't happen again. "I don't understand the reason for this one yet," he says.

Like a terrier who won't let go of Hendrie's pant leg, Fouchard rasps over the speaker phone, "I believe as a precautionary measure——"

Hendrie cuts him off. "I think we had better get— Harold, see if you can get some sort of a better link established . . ."

"Well," says Denton, "people who go up there fall into

a morass; it seems like they are never heard from."

Hendrie realizes that if the confusion is this complete at NRC headquarters, it must be as bad or worse in the governor's office. "We are operating almost totally in the blind," he says. "His information is ambiguous, mine is nonexistent, and—I don't know, it's like a couple of blind men staggering around making decisions . . ."

This has been dragging on now for better than half an hour. Fouchard can't stand it any longer. "Mr. Chairman," he shouts. "Is there anybody who disagrees that we ought to advise the governor what to do?"

They can think of nothing else to talk about. At last Hendrie nibbles the bullet. "Is there a consensus there that we ought to recommend to the governor he move people out within the five-mile quadrant?"

Denton says, "I certainly recommended we do it when we first got the word, commissioner."

Hendrie still isn't sure what to do, but he can put it off no longer. "Let's see if we can get the governor on the phone."

The NRC operators can't get through. The lines are busy. Just then the phone rings on Hendrie's desk. The governor is calling.

"Chairman Hendrie?"

"Governor Thornburgh. Glad to get in touch with you at last. I am here with the commissioners. I must say that the state of our information is not much better than I understand yours is. It appears to us that it would be desirable to suggest that people out in the northeast quadrant within five miles of the plant stay indoors for the next half-hour . . ."

In their time the good citizens of central Pennsylvania have endured fire and flood, revolution and civil war. In all, they have stood their ground, and in time of strife they have dutifully marched to the drum of civil authority. But this is nuts. Stay indoors?

Since Wednesday, they have been hearing about a possible evacuation; now, emergency vehicles are cruising the streets blaring warnings to "close the windows and stay indoors." It's useless to try to check with a cousin or an uncle, since every phone in town if off the hook. And when Met Ed radiation crews appear in the streets with bright-yellow geiger counters, it is the end of the line.

The local bankers are a step ahead; pressed by the surge of withdrawals, most of them shut down within the first hour of operations. The gas lines are already around the corner.

And out on the island itself, radiation specialist Dave Ethridge is having his own second thoughts. His house, his wife, his newborn baby are less two blocks from here. If there were any windows in this goddamn building, he would be able to look across the river right into his living room. He tries to call home but the line is busy. So he uses the plant tie line to Philadelphia and reaches his wife by long distance. "I'm okay," he says, "but the thing is, they might call an evacuation, and the best thing is to grab the baby and pack a bag and leave right now. Don't hesitate."

Ethridge has a vision of some maniac with a shotgun who needs a car, or some panicked kid who runs a red light at a hundred miles an hour. His vision is not that far off. At a saloon in town, a steel-plant worker knocks back a shot and confesses, "I got my piece in the trunk. If the turnpike gets jammed, I'll shoot my way out."

In the village square, a New York reporter stops his car to ask directions, but the streets are empty. The only thing missing from the scene is drifting tumbleweed. Like the reporter, most of the people still in town are headed for the American Legion Hall to see if they can find out anything at Met Ed's morning briefing.

The Middletown Legion Hall is really a bum room with a bar, and in a space that would normally hold a couple of dozen bingo players they have managed to jam a hundred frightened reporters. All three networks are here in force, and shoehorned into the crowd is every TV camera crew from Pittsburgh to Philadelphia.

This thing was scheduled to start at ten o'clock, but so far none of the company officials have been heard from; now there's even money they won't show at all. Then one of the civil-defense trucks passes by and even above the din they can all hear it.

". . . PLEASE CLOSE YOUR WINDOWS AND STAY INDOORS UNTIL FURTHER NOTICE . . ."

Nobody knows whether to bolt for the door or fight for a better camera angle. Some reporters are calling the home office demanding hazard pay; others are trying to arrange for charter flights out of here. Then somebody shouts for attention.

"Ladies and gentlemen, Mr. Herbein just left the site and he should be here any minute." In the end, scoop mentality prevails, and the ladies and gentlemen of the fourth estate make a spectacle of themselves jockeying for position.

After a disastrous false start, Jack Herbein has managed to make a little headway with the working press. The engineering v.-p. was inadvertently drafted as the company spokesman, and over the past two days he has managed to alienate almost everybody from the governor to Jimmy Breslin; but after the hostile briefing yesterday, Herbein took a bunch of reporters aside and gave them a chalk talk on nuclear-power plants. It was exactly the kind of background information that had been missing. The newsmen were grateful, some were even complimentary. So, the man is completely unprepared for the snake pit he steps into this morning.

The people in this room, unlike Herbein, have been listening to the radio. And unlike Herbein, they already know there has been a reading of 1,200 millirem over the plant. So when the beleaguered vice-president steps to the podium and starts talking about readings in the range of 300 millirems, the scene gets ugly.

Altogether, the people in this room have probably witnessed more death and destruction than occurred during the Spanish Inquisition; now they suddenly find themselves scared out of their wits by something they can't see, touch, feel, or taste. They're in no mood for jive. The questioning turns hostile and sarcastic. Herbein can't figure out what's gone wrong.

"We don't see any reason for emergency procedures," he says. "The radiation releases from the plant are less than that of a dental X ray." Clearly, he is the only one in the room who believes this. Everyone else, from CBS to *Rolling Stone*, is convinced the man is covering up.

Somebody wants to know why the company covered up the dumping of waste water into the Susquehanna yesterday. "Why weren't we told about that for ten hours?"

Herbein has had it. He's been on his feet for three days and he's tired of being badgered by the press when he should be back at the plant taking care of business. "I don't know why we need to tell you each and every thing we do," he barks.

Standing behind Herbein, the company PR man bites his tongue and stares at the ceiling, visions of lead stories dancing in his head. Too late now. This is live.

Just up the street from the White House in Joe Hendrie's private office, the five men who run the Nuclear Regulatory Commission continue to dance around the question of evacuation. Every time they are close to a decision, somebody manages to ask one more question. But the good news they are waiting for never comes.

Through all this, several senior staff men in Bethesda keep pushing the commissioners to start moving people. And through it all, Chairman Hendrie manages not to deal with it. Right now, he is out of the room as Ed Case rattles off the latest anxieties. "Well, the concern is that you have got a very damaged core, and it is going to be subjected to a lot of forces hydraulically . . . The core is in a mode that this is just not designed for."

"Wait a minute, Ed, Joe has come in."

Hendrie emerges from a private office and rushes across to the speaker phone on the desk. "Ed, is Harold there?"

"Just a second."

"Harold?"

"I'm here," says Denton.

"This is the chairman again. The president just called over. I think you had better go down to the site. He'd like to see a senior officer and I think you are it."

By now the White House has come to realize what the rest of the country has suspected for the past two days: the situation in Harrisburg is out of control. It's become clear to Jimmy Carter that this collection of Ph.D.'s who run the NRC are incapable of managing this crisis. The president wants a commander on the spot. Harold Denton will be his personal deputy.

Of the five presidental appointees in Hendrie's office this morning, Peter Bradford is the only one with any fundamental doubts about nuclear power. It's not based on anything tangible; he's a lawyer, not an engineer; he was appointed to the commission only eighteen months ago and he's still learning his job. So his anxiety is merely intuitive. And that's hardly an argument to challenge Joe Hendrie's years of experience.

But experience can sometimes be a trap. Instinct, on the other hand, is the subconscious analysis of everything you know, not just the part you choose to remember. Artists, unencumbered by convention, sometimes see things more clearly than the experts.

Bradford's wife, Mary, has the soul of an artist, and she said something at breakfast that still rings in his ears. He was talking about the possibility of evacuation and she said, "What would you do if it were your children, Peter?" It was certainly easy enough for him to picture. His first wife and two kids live in the shadow of the Maine Yankee nuclear plant. He realized he wouldn't want them anywhere near Three Mile Island.

After watching the chairman vacillate endlessly, Bradford pulled him aside at one point and asked him, "What would we do if we had a good friend, and his pregnant wife and small children were in Middletown, and we weren't commissioners?"

Bradford stands nearly six feet eight inches, and Hendrie has to look up to him, whether he wants to or not. Something about Mary Bradford's logical question seems to jolt Hendrie into focus. Again he slips into his private office, this time to call the governor.

At this moment, Thornburgh and Scranton are bouncing off the walls of the governor's office; a couple of minutes ago some overanxious fireman tripped one of the civil-defense sirens and panic is rippling through downtown Harrisburg. Some state employees are already running for the parking lots. Scranton is on the phone shouting at Colonel Henderson, "I don't care who did it! Cut it off!"

Just then the buzzer interrupts.

"Chairman Hendrie calling."

Thornburgh gets on the line. "Hello, chairman. This is the governor."

Hendrie apologizes for the confusion earlier this morning. He says Harold Denton is on his way to the site with the full authority of the NRC and the White House. Thornburgh is glad to hear this, but he wants to know if the commission has come to any decision about moving people. Hendrie doesn't answer the question directly. Instead he uses Bradford's answer. "If my wife were pregnant and I had small children in the area, I would get them out, because we don't know what is going to happen."

". . . Preserve me, O God;
        For in thee do I put my trust . . ."

On the other hand, Mrs. Ed Hauser knows that the Lord helps those who help themselves. Her mind races as she listens to the governor's latest broadcast.

"I am advising those who may be particularly susceptible to the effects of radiation, that is, pregnant women and preschool-age children, to leave the area within a five-mile radius of the Three Mile Island facility until further notice . . ."

She packs the kids in the car and heads for her mother's place in Hershey.

Roger Mattson has taken a different slice at this problem and has come up with some startling conclusions. Yesterday Vic Stello asked him simply to believe the instruments— assume the gauges were not lying after all, but telling the truth. Mattson has been working on the calculations for nearly twenty-four hours, and the puzzle is beginning to come together.

If, for example, you believe that the core might actually be as hot as those thermocouples indicate, then it must have been uncovered. That would account for the astonishing radiation levels. At those temperatures, the zirconium fuel pins must have literally burned up—oxidized—which would mean that an enormous amount of hydrogen (H) must have been freed from the water (H_2O) by the chemical reaction. If you assume that the coolant system is full of hydrogen, this explains why the pipes are blocked in the high spots. Each new snippet of data that reaches Mattson drives another rivet into the coffin. By noon on Friday the bearded scientist has a picture of the situation in the bowels of the plant that is much more frightening than anyone imagined.

For several hours now, he has had the agency's top systems experts searching the Three Mile Island blueprints to see if there is some pipe, some valve, some tricky way to get that gas out of there so that they can start the water flowing again. But when they report back to him, the answer is negative. For a long moment, Mattson stares at the piping diagrams. It is hard for him to realize that with all the miles of plumbing and tons of hardware, there isn't some way to get a little gas out of that system. But there isn't.

He is defeated, overwhelmed by the sense of loss. His life has been dedicated to the prevention of this moment. Now here they are, teetering on the brink. He scoops up the papers and heads for the Incident Response Center.

The emergency-management team is huddled together when Mattson bursts in. "I need to talk to you people," he says. "Sit still for a minute and let me tell you how bad this is." Several of the men sink to their chairs as Mattson fills them in. "I don't know what decisions you're making, but if you're counting on that reactor and some miracle of engineering to get you out of this, it ain't gonna happen." In brutal language, he paints a picture of the core that leaves them bug-eyed.

Then he drives home the linchpin. "You people have a responsibility. There's no way out of this situation and it's on the precipice. You better be doing something with the people."

They are stupefied. How can this be? ". . . the radiation levels are still below the EPA guidelines . . ."

"Jesus Christ," says Mattson, "you don't take emergency action *after* the doses are lethal off-site. You don't wait until the core is melting out the bottom of the plant."

Denton's deputy, Ed Case, has been recommending evacuation since nine o'clock this morning, but he hasn't been able to get Joe Hendrie off the dime. He shoves Mattson in front of the speaker phone and puts him on the line to the commissioners.

Like a locomotive, Mattson can be direct. He tells Chairman Hendrie, "My best guess is that the core uncovered, stayed uncovered for long periods of time. We saw failure modes the like of which have never been analyzed."

Hendrie is slightly built, gaunt and weary. Now he seems to be shrinking in his chair. But the voice on the speaker phone continues a depressing trip-hammer litany.

"We just learned—I don't know—three hours ago," says Mattson, "that on the afternoon of the first day, some ten hours into the transient, there was a twenty-eight pound containment-pressure spike. We are guessing that may have been a hydrogen explosion. They, for some reason, never reported it until this morning. That would have given us a clue hours ago that the thermocouples were right and we had a partially disassembled core."

Then Mattson gives him the bad news: there is an immense bubble of hydrogen trapped right above the core. "They can't get rid of the bubble," he says. "They have tried cycling and pressurizing and depressurizing; they tried natural convection a couple of days ago; they have been on forced circulation; they have steamed out the pressurizer; they have liquided out the pressurizer. The bubble stays."

It's the ultimate dilemma. In order to shut the reactor

down, they must reduce the pressure. But when they lower the pressure, the bubble gets bigger. According to Mattson's calculations, if they reduce the pressure enough to do any good, the bubble will push all the water out of the core.

"Yes," says Hendrie. "You are going to blow right down and empty the core."

At that point the uranium fuel pellets will begin to heat up at the rate of three degrees a second. The crucial question is whether they will be able to blow the hydrogen out of the core—and get water back in there—before the core begins to melt. If they can't complete a blowdown in under twenty-five minutes, they might as well head for the hills.

"I've got a horse race there," says Mattson. "Do we win the horse race or do we lose the horse race?"

"Suppose we try?" says the chairman.

There is a long pause. "Well . . . if we don't know any more than we know right now . . . I think you're going to lose the horse race." Unlike Hendrie, Mattson understands how lucky they've been so far. "The latest burst didn't hurt many people," he says. But luck can change. "I'm not sure why you are not moving people. Got to say it. I have been saying it down here. I don't know what you are protecting at this point. I think we ought to be moving people."

This advice, clear and simple, from one of the agency's top safety experts, is incredibly depressing, but not all of the commissioners manage to hear it. Bradford and Gilinsky, the two cautious members who might be counted on to push for evacuation, are both out of the room. Gilinsky is at his desk returning calls. Bradford is trying to reach his son up in Maine to explain why they won't be able to go skiing this weekend.

But Commissioner Kennedy is here. "How far out?" he asks.

"You aren't going to kill any people out to ten miles. There aren't that many people and these people have had two days to get ready and prepare."

"Ten miles is Harrisburg," Kennedy reminds him.

Hendrie is confused, stunned, like a weary boxer. "I don't know Roger, you . . ."

There is a long pause. Mattson tries again. "It's too little information, too late, unfortunately," he says, "and it is the same way every partial core meltdown has gone. People haven't believed the intrumentation as they went along. It took us until midnight last night to convince anybody that those goddamned temperature measurements meant something."

Hendrie decides to deal with this later. "Okay," he says, "get back on the phone with them and, if you will, keep us posted as you go along."

And that seems to be that. From here, Hendrie and his two colleagues turn their attention to other business. Someone has suggested that the NRC should set up a press center out in Bethesda. "Okay," says the chairman, "I think on balance that's a good idea."

The president of the United States is one of the few people on the planet with the ability to turn thought into instant action. When he talked to Governor Thornburgh this morning, Jimmy Carter could see that the communications problem is a major factor in the chaos. The phone system has broken down. Dial a number in the 717 area and you get a busy signal right after the area code.

This, at least, is something Carter can fix. Signal-corps technicians are already in the air, bound for Harrisburg. In less than two hours, there will be dedicated signal-drop lines connecting the White House switchboard with every player in the game. And in the offices of the NRC, the governor of Pennsylvania, and the control room at Unit Two, hot-line telephones appear like magic.

The White House, by way of the Pentagon, has also arranged transportation for Harold Denton and his staff. An air-force chopper is inbound for the pad at Bethesda Naval Hospital, less than five minutes from NRC headquarters. Fortunately, Denton has a bag packed; he was supposed to go to Phoenix two days ago. But Harold Denton has one last thing to do before he leaves for the site, one last thought to pass along to the senior staff.

He finds Vic Stello and Roger Mattson and pulls them aside. Denton ushers them into a room. He closes the door behind. He turns to them with fire in his eye. "This technology," he says, "has relied upon the safety systems that you two have played a large part in. *Don't tell me that those goddamned safety systems don't work.* You find a way to get out of this situation."

He picks up his gear and leaves for Three Mile Island.

15

From her front yard, Sarah Schneider has to look up to see the top of the cooling towers; only the highway and the river are between her mailbox and the island. But all this hustle and bustle hasn't upset her. Mrs. Schneider is a Met Ed booster. Her husband, Dewey, has worked for the company as a lineman for nearly twenty years and they've always done right by him. So when Met Ed said they wanted the house, it was okay with Dewey and Sarah.

The Schneiders' two-bedroom bungalow is just across the driveway from the Observation Center. The center has been Jack Herbein's command post for the past three days, but they're out of room over there and the tide of support personnel is rising by the hour. Westinghouse has dispatched a team of engineers from Pittsburgh to see if they can lend a hand. The NRC will have eighty-three inspectors on-site by this evening. Babcock & Wilcox is coming in force. All these people need office space, lab space, drafting tables, desks, toilets, and telephones.

Out back, Sarah can hear a bulldozer working in the hayfield—they're leveling a pad for the helicopters—and workmen are running power and phone lines to the rows of house trailers being set up in the parking lot.

Slowly the noise of the construction equipment is

drowned by the thunder of an air-force helicopter approaching from the south. This machine is much bigger than anything that has landed so far, and as it settles, a whirlwind of blinding dust sends the hard hats running for cover. From the cloud emerge Harold Denton and nineteen assistants.

As soon as they are clear of the blades, Denton and his people are hustled across the lot in through the back door of the Schneider house. Denton passes the laundry, which is now a darkroom, and goes through the kitchen, where Sarah is serving coffee and sandwiches. He is polite, considerate; he introduces himself to Mrs. Schneider and apologizes for all this.

The phone rings and it's for Denton. He's impressed. Who could have tracked him here so quickly?

"Mr. Harold Denton?"

"Yes."

"This is the White House calling. One moment, please, for the president."

Denton's temples barely have a chance to start pounding before he hears the familiar southern accent. Carter comes right to the point. Like most of these nuclear people, he's an old navy man. He understands that the commander on the spot must have absolute authority. Denton is to be the presidential deputy, a technological ambassador without portfolio, backed by the Great Seal of the United States. Anything he needs will be at his disposal. Pick up the phone; call Jack Watson; tell him what, when, and where. Finally, the president warns Denton to tell it like it is. "Don't overplay it or underplay it," he says. "And as soon as you find out what's going on, call me back."

Denton hangs up and turns to PR man Joe Fouchard. "Wow."

Unlike the collegial body of academicians down on H Street, Harold Denton, head of the NRC technical staff, is used to running an organization and issuing orders. At last, a battle commander has hoisted his flag on the bridge. He is the president's man. If anything goes wrong now, it will surely be his ass. Immediately, Denton gets word through to the control room over on that godforsaken island: DON'T TOUCH ANYTHING UNTIL YOU CHECK WITH ME.

Fifty people are jammed into Sarah Schneider's tiny living room. They do their best not to break anything as the Met Ed senior staff briefs Denton on what they know. Although it is a staggering litany of destruction and danger, Denton is relieved. He was afraid the Met Ed people were still living in fantasy land; at least they have come to the realization that the situation is potentially catastrophic.

Again, Denton warns them not to make the slightest adjustment without his express permission. He divides into four groups the hand-picked crew he brought with him and sends them over to the island to check the facts. Then he ducks out the back door and into the state-police communications trailer to call the White House.

Dr. Henry Myers is a physicist on the staff of the House Committee on Energy and the Environment. He once worked for the NRC and is aware of their tendency to emphasize the bright side. As the staff aide responsible for knowing what's going on in the NRC, Myers calls NRC headquarters to see if he can get any sense of what's really happening. A colleague is standing by Myers's desk as he talks to the man in Bethesda. Holding the phone with his shoulder, Myers pulls out a piece of typing paper and writes a single word:

BAD

Roger Mattson is still on the line to commission headquarters, trying to nudge this august body toward some kind of

decision about evacuation. He feels like a man on fire. Harold Denton's parting words still ring in his ears. As much
as any man alive, Mattson feels responsible for this. True,
there were a thousand fingers in the pie; but Mattson super-
vised the safety regulations. He and most of his colleagues
had come to the conclusion that nuclear power was a mature
technology; any new discoveries would simply be improve-
ments. Now his mature technology has suddenly developed
infantile tendencies.

The leak isn't big enough; that's the problem, simply
stated. If they had a major hole—if there were a great, gush-
ing, four-foot rip around one of those humongous coolant
pipes—then they could handle it. The trapped hydrogen would blow out in a minute and they could flood the
core with emergency cooling water. The machinery is designed to deal with a major break. Mattson made sure of
that. But they don't have a major break. Right now the only
hole they can open up is that little tiny goddamn valve
on top of the pressurizer. Chairman Hendrie suggests
that they try to break something intentionally. "Can you
pick out a four-inch pipe and put us in a mode we understand?"

But Mattson has already been through this. Incredible as
it seems, in all that labyrinth inside the containment, in all
those miles of interconnected pipe, there is not one single,
piddling four-inch bleed valve that can be opened from the
control room. They designed the system to be leak proof;
now they are trapped by their own efficiency.

"They are working on some alternates," says Mattson. "B
& W wants to start up all reactor coolant pumps, burn them
out, blow the seals, and hope they cause a loss of coolant
accident that way . . ." This idea from the plant's designers
is a little too Draconian for Mattson. What if they need those
pumps later? "It's a failure mode that has never been studied," he says. "It's just unbelievable . . . I don't know what

I would say if a reporter called me and asked me if we had melted fuel. There are some things going on that we can't explain unless we did."

"What is your principal concern right at this minute?" asks Commissioner Gilinsky.

"Well, my principal concern is that we have got an accident that we have never been designed to accommodate, and it's—in the best estimate—deteriorating slowly, and the most pessimistic estimate—it is on the threshold of turning bad." Again he tries to build a fire under them. I don't know what you are protecting by not moving people."

Gilinsky is impressed. He wants to do something, but he's not sure what. Chariman Joe Hendrie isn't here right now. He's over at the White House situation room briefing the president. "Suppose one did say right now that we ought to execute evacuation," says Gilinsky. "Are there plans that would be put into effect, or what would happen?"

Somebody tells him that evacuation would be handled by the state of Pennsylvania. So Gilinsky suggests that they call the state and find out if they are ready to move, just in case. To Commissioner Dick Kennedy, however, the mere mention of the word "evacuation" is a white flag of surrender. Kennedy's background is military, and in a tight spot, he's inclined to tough it out.

"I don't think we ought to do anything like that until the chairman comes back," he says. "First of all, the minute you do that, the press is going to read that as a signal of a decision." Kennedy has a mistrust of the press; he knows how they sensationalize these things.

But Gilinsky says he just wants to check with somebody at the state level to find out if they are ready to move. At least the commissioners should know how long it will take to get an evacuation underway. Kennedy thinks Gilinsky is naive. "It's going to be in the newspapers this evening at five o'clock 'NRC CONTEMPLATING EVACUATION.'

If that's what you want, all right."

As Mattson listens out in Bethesda, this argument between the commissioners drags on. Finally, he motions to the agency's evacuation expert, Doc Collins. Maybe Collins can tell the commissioners what they want to know without calling Pennsylvania or alarming the press. Collins comes to the phone and says that yes indeed, the state of Pennsylvania does have an evacuation plan and the governor can order an evacuation any time he wants to.

"How long would it take?" asks Gilinsky.

Collins rumples his map. "Let's see . . . Harrisburg is in which county? Yes, there are a lot of little towns around there, too, but the sector would be, you know, seventy degrees wide. I would say they ought to be able to get all the small towns in the counties and the local folks out within an hour, and probably certainly have the city cleared by two, or you know, something in that order."

Fortunately for Doc Collins, the governor of Pennsylvania can't hear this wild speculation about moving 150,000 people in two hours. So Commissioner Gilinsky and the others are reassured. They can wait till the last possible instant before making a decision.

Out in Bethesda, just up the stairs from Roger Mattson's temporary office, the newly manned NRC press center is doing brisk business. Dudley Thompson, one of the senior staff executives, is giving an update on the accident to a mob of reporters when somebody repeats the question everybody has been asking. "Is there a chance of a meltdown?" By now, most of these people have seen *The China Syndrome*, and in the movie, "meltdown" is the ultimate catastrophe.

"Yes," says Thompson, until they get rid of the hydrogen bubble, "there is the possibility of a meltdown."

Of course there is always the possibility of a meltdown,

and this has been stated before. Joe Hendrie himself said it
to Congress yesterday. But Hendrie was quick to add that
the possibility is very remote. This time, maybe because
Thompson is aware of the furor raging among his colleagues
downstairs, he fails to emphasize the long odds. In any
event, something in the way he says it galvanizes the press.

At the state house in Harrisburg, the governor has just
said good-bye to Chairman Joe Hendrie after another reassuring call when the teleprinters in Paul Critchlow's office
begin to chatter. Everybody hears the bells on the UPI
machine; they know it's going to be bad news. Critchlow's
assistant rips off the copy and hands it to him.

> 03 30 79 BETHESDA MD 1602
> ADD NUCLEAR
> NRC ADMITS ULTIMATE RISK OF MELTDOWN.

The phone rings in the governor's office; Harold Denton
is calling to introduce himself.

Before Dick Thornburgh was elected to this office, he was
a federal prosecutor. A lot of people are behind bars at this
very moment because of him. Now Harold Denton is in the
witness chair, and Thornburgh wants the truth, the whole
truth, and nothing but the truth. What's this meltdown
business?

Fortunately for everybody, Denton has a quiet, low-key
manner that inspires confidence. He sounds knowledgeable
without displaying the technological arrogance of a man like
Jack Herbein or Vic Stello. He talks like a Sunday-school
teacher. "Yes," Denton admits, "there is the possibility of a
meltdown." But it is very remote.

Denton explains that he has divided his team into four
task forces and they are due to report back from the island
within the hour. As soon as that happens, he will come to
Harrisburg to brief the governor and the press. In the mean-

time, he has good news and bad news. The good news is that Met Ed seems to have a better understanding of the situation than he thought. The bad news is that the uranium fuel is severely damaged. And he confirms that there is an immense bubble of hydrogen immediately above the core.

In a corner of the Governor's office, state radiation expert Tom Gerusky buries his head in his hands. "Oh, my God," he says.

Fouchard peeks through the drapes. Out on the lawn, Sarah Schneider's flowers have given in to the trample of reporters. Everybody saw that helicopter land and they want to know who these people are and what's going on. Obviously, Denton is going to have to face this mob or he'll never get out of here.

They have come from all over the world, and in the taverns of Harrisburg it's no longer unusual to hear Japanese, German and French. About three hundred reporters and photographers are here already, and more are on the way. The *Philadelphia Inquirer* alone has sixty people working on the story.

The state capitol is first stop for most new arrivals, and there has been such a tide of reporters up the marble steps that the police have simply stopped checking credentials. "They told me that if they don't look like derelicts to let 'em in," says the guard. "But some of 'em do look like derelicts."

One of New York's most famous derelicts, Jimmy Breslin, is already on the case.

I kept listening to the car radio tell about the malfunction in a nuclear plant near Harrisburg, Pa., yesterday. There was a broken valve somewhere in the nuclear reactor and this caused, the power-company man announced, "a puff of radioactivity" to be released into the atmosphere. I loved that. "A puff."

That's what they hit the east ward of Nagasaki with. "A puff."
"Step on it Dennis," I yelled, "it could be the end of Pennsylvania."

My heart jumped when I heard the lieutenant governor of
Pennsylvania say over the radio, "The incident is minor. There
is and was no danger to public health and safety."

"That means they must have thousands dead," I yelled. "Hit
it, Dennis."

This skeptical view of the official word has become the
operating assumption. Nobody believes anything Jack Herbein says anymore; his brief career as a PR man is about
over. The governor has admitted that nobody knows for
sure what's going on. And the NRC has a different story
every time you talk to them.

It's a gig everybody hates. There's no action; just a war
of nerves. The photographers are drinking heavily. How
many times can you shoot a picture of a cooling tower? At
Lombardo's, they keep their spirits up with gallows humor:
"What's the weather tomorrow?"

"Partly cloudy, with a 40 percent chance of survival."

Unlike in a war, there's no front line, no rear area. Like
smoke, radioactive gas travels in wisps and eddies. You can
be standing on one side of the street and get nothing, while
the guy on the other side gets a dose. Unlike smoke, this stuff
is invisible, and that's what makes it so creepy. It can zap you
in your motel room while you're watching Johnny Carson,
and you might not know it for years.

The first ones on the scene were the old state house hands
like Jim Panyard of the *Bulletin*. He always felt that politicians were masters of obfuscation; he had no idea about
engineers. "All my sources are speaking a foreign language," he says: "millirems, manrems, rads, and picocu-
ries . . ."

Most of the reporters on this assignment are as much at

sea about millirems as Jim Panyard is. Many of these journal-
ists are hearing words like this for the first time. But sprin-
kled here and there among the press corps are a few technological sleuths who have been on the trail of this story for
years.

David Burnham of the *New York Times* is not surprised
in the slightest by the events of the last two days. He's been
expecting an accident like this. He has written column after
column about various nuclear close calls, but Burnham's
often pessimistic articles were usually buried among the
Bloomingdale's advertisements. But he persists. Burnham
was the reporter who waited that night in an Oklahoma
motel room when Karen Silkwood failed to show up with
the evidence against Kerr-McGee.

Tonight, Burnham is on the trail of a very hot lead.
Through one of his sources—a former NRC engineer—
Burnham has heard about an inspector out in the Chicago
regional office who was trying to blow the whistle on the
Babcock & Wilcox plants for the last year and a half, but
nobody would listen.

Burnham scratches the number on an note pad and dials
Chicago direct.

"Hello."

"Jim Creswell?"

"Yes."

Burnham introduces himself. "It looks like you won," he
says.

There is a long silence. "It's a hell of a way to prove your
point," says Creswell.

16

Saturday, March 31

In the small hours of the morning, the intense radiation inside the reactor building is beginning to take its toll. Radiation can be as devastating to machinery, particularly electronic gear, as it can to human beings. The radiation level in there has been God knows what for the last three days, and some of the insulation on the electrical wiring is starting to go. They are losing instruments. As time wears on, they will know less and less about what's happening.

At two A.M., a party of radiation technicians is dispatched from the control room to the equipment hatch of the reactor building to see if they can get some direct indication of the activity level. Cautiously, they descend the steel stairway in the service building down to level 305 and enter the corridor that leads to the containment. The hatch is an immense double-steel airlock twenty-four feet in diameter. Peeking around the corner, they extend the Teletector probe a full twelve feet and touch it to the surface. The needle jumps to sixty rem per hour.

This is not radioactivity leaking through the hatch. This is radioactivity inside the containment, shining through several inches of armor plate. If you assume this great door cuts the radiation by a factor of ten, that means the level inside is 600 rem per hour.

The men back away and quickly retreat down the concrete corridor. It will be months—maybe years—before they will be able to open that hatch. And this is heart-stopping news for Met Ed president Walter Creitz.

At the moment, the press corps has an image of Walter Creitz as a defensive corporate lackey trying to cover up a major disaster. Actually, he's a rather pleasant fellow who struggled up the company ladder the hard way, and who never imagined in his most chilling nightmare anything like this. He could not have been less prepared for the second coming of Christ.

Now he can see his career—indeed, his company—slipping under. How could this have happened? Just last week —well, that's pointless. It's happened. But as he watches the sun rise over his mortally wounded plant, Creitz must surely have an impulse to weep.

The designers, the operations people, all those technical experts—everybody always said the safety systems would prevent this. Now . . . a billion-dollar investment, virtually wiped out by the turn of a valve.

And there is Unit One, a few hundred yards to the north, waiting for them to hit the switch. It sits there, completely undamaged, fuel loaded, ready to go. But they don't dare turn it on. The Unit One systems have to be available to back up the failing equipment in Unit Two. Together, these plants are losing two million dollars a day. The interest— just the interest—on the money borrowed to build the two reactors is $160 million a year.

Like a cockleshell before the hurricane, Creitz is being driven onto the rocks by wave upon wave of unforseeable disaster. He's done his best. He and Jack Herbein tried to give people an understanding of the situation. Perhaps at the beginning they did tend to look on the bright side; they were trying to calm everybody down. But there is no bright side to this thing. It threatens to eat them all.

From the minute Harold Denton got here, Creitz has been trying to get the NRC to close ranks with the company. Together, he felt, they could limit some of the damage Met Ed is suffering in the press. So Creitz approached Denton as soon as he got off the helicopter and asked him to authorize a joint press release. Fouchard read it and nearly had a seizure; he told Creitz to forget it. The last thing he wants is to blow Denton's credibility by trying to save Met Ed. It's too late for that.

But in the service of his company, Creitz is good for one more try. This morning he calls on the NRC command post in Sarah Schneider's living room; he suggests to Joe Fouchard that the NRC and Met Ed hold a joint press conference. Fouchard will have none of it. He won't even let Met Ed use the same meeting hall as the NRC. Exhaused, beaten, Walter Creitz heads for the Legion Hall basement, and at eleven A.M., he steps into the international spotlight one last time to announce that this will be the final Met Ed press conference. From now on, he says, Harold Denton at the NRC will be the sole source of information.

But Jack Herbein is there for a final go at the reporters (a species of human he still can't fathom). He has changed into a more photogenic suit—something a little less glossy —but he still hasn't learned anything about the information business. After three days of having his words shoved down his throat, he is still inclined to optimism. He says his people are slowly getting rid of the hydrogen, and as far as he's concerned, "the crisis is over."

The press reaction is vicious. Some of the younger reporters climb on tables and shout questions with such sarcasm and hostility that it's obvious they're not after answers; they're after blood. Creitz is bewildered. How could they have gone from being a small, respected utility to being public enemy number one in seventy-two hours?

Joe Hendrie has made plenty of mistakes on behalf of the NRC over the last three days, but sending Harold Denton to Harrisburg was not one of them. It turns out that Harold Denton is the man for the job. He alone seems to understand that the only way out of this swamp is to communicate with the press in plain English. Early in the accident, he was grilled by a reporter and tried to explain his position by saying, "If you knew what I know, you would think like I think."

"Well," said the man, "what do you know?"

So when a newsman asks him to comment on Jack Herbein's statement that the crisis is over, Denton says, "The crisis is not over."

At the moment, he is onstage at the Middletown Borough Hall, a gymnasium in the center of town that Joe Fouchard has turned into an instant press center. The place is filled with reporters fresh from the Met Ed briefing over at the Legion, and Denton tells them—contrary to what Herbein said—that the hydrogen bubble has not been reduced in size. It's as big as ever. Since this is what everybody believes anyway, Denton establishes himself as a straight-talking realist.

But he does not bother to mention his other grave concern. Last night, it occurred to him that Met Ed might not be equal to the occasion. It's not so much a question of competence as "technical depth." Met Ed is a small company, and they have been manning the pumps without letup for three days. Denton is concerned about the ability of these exhausted men to make careful decisions.

At one point, he discusses with his staff the feasibility of having the NRC take over the plant. Buried in the original Atomic Energy Act is a clause that gives the commission authority to take possession of plants in an emergency. But it is never a serious consideration. The truth is that the NRC does not have the operations people to actually run a plant like this even if they wanted to.

In his next call to the president, Denton confesses that he's uneasy about Met Ed. He suggests that the White House put out a call for help. Carter gets his assistants on the line and Denton ticks off a list of the principal U.S. companies in the nuclear business. Within the hour, corporate jets all over the country are rolling out and fueling, as top technical managers from Westinghouse, General Electric, Combustion Engineering, Commonwealth Edison, and Duke Power Company prepare for the run to Harrisburg.

By now the second-level staffers in the White House are trying to leapfrog ahead of the situation. Gene Eidenberg, one of Jack Watson's aides, wants to make sure nobody runs out of gas in the middle of an evacuation. Last night, he had the Federal Disaster Assistance Administration move 200,-000 gallons of gasoline into the Harrisburg area to top up the key filling stations on the roads out of town.

Admiral Rickover and the navy have been helping all along, of course; the air force is flying in weather-mapping gear; and the army is trucking communications equipment from Philadelphia. Both the army and the air force are involved in the effort to rush a million pounds of lead bricks to the island. A convoy of nine army trucks is bringing every lead brick they could spare from the National Bureau of Standards, and flights of C-141 cargo jets are on the way with lead bricks from Brookhaven National Laboratories. The bricks will be used to build shield walls around essential equipment, near the containment, that is now unapproachable because of high radiation.

And from Buena, New Jersey, an army semitrailer is rolling down the Interstate with a load of medicine droppers. They are for the potassium iodide. A dose of potassium iodide fills the human thyroid with iodine and helps prevent takeup of radioactive iodine. Mallinckrodt Chemical Company in St. Louis is working around the clock to produce a million doses.

Yesterday Governor Thornburgh raised a question that has been eating at Joe Hendrie ever since. Thornburgh asked if there were any possibility the hydrogen bubble could explode. Without hesitation, Hendrie assured him that the chances of an explosion were close to zero. But as soon as he hung up, Hendrie began to wonder: what if they had all missed something? From that moment on, the thought hasn't left his mind.

Hendrie knows there has to be some oxygen inside the reactor vessel. For one thing, the intense radiation is splitting apart the water molecules into hydrogen and oxygen. Nobody believes this is happening fast enough to worry about; on the other hand, you don't need that much oxygen in order to have a problem. "You need to get up to 4-percent-by-volume oxygen to have a mixture which is minimally flammable," says Hendrie. "Above that, you get to levels at which the ignition is progressively easier and detonation velocities are faster and you get more bang out of it."

How much more bang? Even a small explosion could collapse what's left of the core; a large one could turn the reactor vessel into a grenade. Hendrie does some back-of-the-envelope calculations; the answers aren't that encouraging.

Roger Mattson is already on the case. Hendrie asked him to start recalculating the estimates as soon as he felt that first clutch of anxiety. The chairman's sense of urgency was contagious, and Mattson has turned loose an impressive array of talent on the problem.

A couple of years ago, somebody in the NRC said the agency should fund some basic research into hydrogen to determine how the gas would behave in odd situations like this. Management rejected the idea. Now the research will be done overnight in a dozen universities and laboratories, and almost every scientist in the country who ever had a word to say on the subject of hydrogen will be drafted into Roger Mattson's "bubble squad."

Basically, Mattson has set up two separate groups following two lines of inquiry. One group is trying to determine if and when the bubble will be explosive. The other is trying to estimate the effect of such an explosion; this is a much simpler problem, so these answers come in first. The "effects" people say that an explosion inside the vessel might generate a shock wave of 14,000 pounds per square inch. This is bad news. Babcock & Wilcox says the head bolts will fail at 11,000 pounds.

If the lid blows off the top of the reactor vessel, high-velocity shrapnel might breach the containment and several billion curies of radioactive debris could be instantly released to the atmosphere. Small wonder Hendrie and the others are concerned about this scenario. It has roughly the same fallout potential as a one-megaton bomb.

In this rather unsettled frame of mind, Joe Hendrie and the other four commissioners leave for the Incident Response Center out in Bethesda. The PR people said the place was crawling with reporters and they felt it would be a good idea for the chairman to show up and put the proper face on things for the weekend editions.

This mid-afternoon briefing seems to go well enough, but Hendrie's anxiety seeps through. He goes into too much detail, and his speculation about the future is more ominous than anything the commission has said so far. When a reporter asks about the worst case, the chairman admits that when the operators try to reduce the hydrogen bubble, they might order a precautionary evacuation as far as twenty miles out. Hendrie doesn't realize it, but several reporters are quite impressed by some of the facts that have slipped out in this rambling discussion.

The four other commissioners have come to Bethesda with the chairman, so they will all be together if an emergency decision is required. Their worst fear is that the next call will be some mathematician from Stanford with word that the hydrogen bubble is already explosive. After the press conference, the five men gravitate to Roger Mattson's desk for reassurance. But once again Mattson is as blunt as a pile driver.

"Let me say as frankly as I know how, bringing this plant down is risky," he says. "No plant has ever been in this condition, no plant has ever been tested in this condition, no plant has ever been analyzed in this condition in the history of this program, and there is risk in doing that in short order with a damaged core." Mattson warns them that the moment is fast approaching when they will have to make a decision about which path to follow.

Commissioner Kennedy says, "Well, you don't want to rush that. But neither do you want to sit there . . ."

Mattson is maddened by this indecisiveness. They are not talking about weeks. "It could be as short as a couple of days," he says. "Your pump is running in a condition it doesn't like to run in . . . You've got radiation in that containment—with equipment you're depending on now—that's not radiation qualified. It's going to reach a time when it's time to do it. The risk of staying there is going to continue to grow."

"All right," says Victor Gilinsky. "One of the things we're going to discuss is should we be recommending anything about moving people?"

Mattson says, "Are you asking me for a recommendation?"

"No, no," says Gilinsky. "I'm trying—no, no, I understand . . . Well, if you have one, I'll take it . . ."

"Let me give you a feel for——"

"No, no, I don't want to push you on that one, Roger ..." Like the rest of the commissioners, Gilinsky now seems eager to avoid the bottom line. As they waver on the brink of a decision, they are rescued once again by Dick Kennedy.

"I keep thinking," says Kennedy, "you know, we may not be as close to the edge of that precipice as it seems all the time to us. Which, as a matter of fact, time has shown to be true. It's that step to the edge of the precipice. The closer you get to it the more comfortable you are, and yet you're quite a long way from the edge of it."

It's been an hour since the press briefing, and the first returns are just coming in. Somebody brings Hendrie the wire-service copy and he pales. "Oh, boy. No matter what you say, the press screws up." He mimicks a TV announcer. " 'The chairman said that when we try to get rid of the bubble we're going to evacuate everybody out to ten or twenty miles.' Oh boy." He can see the headlines. "What I said was, what we told the governor the other day . . . Oh, boy . . ."

"Yeah," Kennedy reminds him, "you said ten to twenty miles."

"Yes, he said that."

"He trimmed it."

"And we qualified it somewhat later."

"Well," says Hendrie, "that was a press dodge, you know; the press people play this game. They now call up the State Emergency Planning Office. They say, 'The chairman of the NRC just said you've got to evacuate. What have you got to say about that?' " Hendrie shakes his head. "Which amendment guarantees freedom of the press? I'm against it."

The hot-line phone rings. It's Thornburgh. Hendrie picks it up. "Governor, how are you? No, I was just getting ready to call you . . ."

When Dick Thornburgh was a federal prosecutor, he

favored limited use of the death penalty. He was the man who put away the hired killers of UMW official Jock Yablonski; he feels there are times when the death penalty is useful. Right now, he's beginning to wonder if it might not take something like the electric chair to get Joe Hendrie's attention. A twenty-mile evacuation involves 600,000 people, something the state is completely unprepared for.

"No, that's not correct," says Hendrie. "I'm afraid that's one of those cases where the press is trying to work you—work us at cross-purposes." The chairman explains that he was talking about a hypothetical situation; the commission does not have any immediate plans for a twenty-mile evacuation—at least not right now.

But after reassuring the governor at some length, Hendrie closes on a note of caution. "Listen," he says, "before I go away, why, the commissioners wanted me just to note—not to be a Gloomy Gus, but we do have to keep in mind that we could need to move, with regard to protective actions for the public, in a hurry, and the alert status of all the emergency teams is a matter of considerable importance, we think."

In the paneled corner suite of the old state capitol, beneath the gaze of William Penn, as the governor and his aides listen to Chairman Hendrie on the speaker phone, everyone in the room can see Dick Thornburgh's anger building like a wave.

Stan Benjamin of the Associated Press is no stranger to the nuclear beat, and like Dave Burnham of the *Times*, he has his own "Deep Throat" inside the NRC. During the press briefing out in Bethesda, Benjamin noticed Joe Hendrie's anxiety about the hydrogen bubble and decided to follow it up.

After a couple of quick calls, he discovers that the forced

calm of the chairman's public posture is indeed a façade. Behind closed doors, Hendrie is directing a frantic investigation into the hydrogen bubble, and several people in the agency now believe it could soon be explosive. One official, who chooses to remain nameless, admits they might have as little as two days.

At 8:02 P.M., the Associated Press moves a story advisory on the "A" wire. In the state house, press aide Paul Critchlow is once again the first to get the word. He hears the bells and steps in to take a look.

URGENT (WITH NUCLEAR)
NRC NOW SAYS GAS BUBBLE ATOP THE NU-
CLEAR REACTOR AT TMI SHOWS SIGNS OF
BECOMING EXPLOSIVE. A STORY UPCOMING.

Critchlow rips the copy out of the machine just as the switchboard lights up. Immediately he puts a call through to Denton down at the site. "What the hell is going on here?" shouts Critchlow. "Is this story accurate?"

"No," says Denton. "It's not accurate." It's out of context, he explains. "Yes, the bubble is potentially explosive, but you are talking a longer period of time than that." He feels it can be dealt with through a variety of measures they are taking. Denton says he'll hop in a car and come right up and try to straighten everything out.

Critchlow storms back into his office and accosts Karl Abraham, the NRC liaison man. "Karl, you should have told your people in Washington to keep their fucking mouths shut, because the governor is getting sick of it. You're causing a panic!"

Critchlow plunks down at his typewriter and begins pounding the keys. Time is of the essence. He doesn't bother to check with anybody. He hammers out a press release over his own signature. He rips it out and heads for

the press room. Just then the doors crash open and two
dozen out-of-town reporters jump him. To hell with the
story; is it time to pull out?

At this moment, Joe Fouchard and Harold Denton arrive
at the capitol. They are wandering through the marble halls
like a couple of refugees when they're spotted by Dick
Lyons of the *New York Times*. "Where are you going?"
demands Lyons.

"I'm looking for the men's room," says Denton.

Fouchard says, "Where is Paul Critchlow?"

"He's in the press room trying to calm the storm."

"What storm?"

"Joe, you don't understand," says Lyons. "People are
checking out of hotels in this town." Suddenly Fouchard
realizes that Lyons himself is highly agitated.

"Let's find Critchlow," says Fouchard. He grabs Denton.
"We'll go in and try to put this thing in perspective right
away."

After a quick stop at the press room to save Critchlow
from being eaten alive, they run to the governor's office.
Denton explains that they have at least nine to twelve days
before the bubble will be explosive. He says that's plenty of
time because the operators have come up with several possibilities for dealing with it. He's confident they aren't fac-
ing imminent catastrophe.

The governor says, "If we're gonna knock this thing,
we've got to knock it now." They jump up and head for the
press room. At eleven P.M., Denton and Thornburgh go
before the live network cameras. Denton is calm, forceful,
reassuring. Amazingly, in just twenty-four hours as a public
figure, he has managed to build enough credibility to halt
the panic in mid step. People are still scared, but no longer
terrified.

Then Thornburgh offers a surprise. As a "vote of confidence in the kind of work that is proceeding there and a

further refutation of the kind of alarmist reaction that has set in in some quarters," he says, President Jimmy Carter will be coming to Three Mile Island, possibly as soon as tomorrow afternoon.

The day is finally ended and Scranton is at home when he gets a call from Kevin Malloy at civil-defense headquarters. Malloy is pissed. Since Wednesday morning, his people have been leaning forward in the trenches, catching a few winks on cots in the office. Most of them have been up for three days, Malloy included. They had access to all the rumors without the resources to follow up. Malloy thinks Scranton has been high-handed. "If we don't get some answers down here, Bill, we're calling an evacuation on our own authority at nine o'clock tomorrow morning."

Scranton clutches his head. "Oh God, Kevin, please don't do this. I'll come down first thing in the morning and talk to you." After a couple of minutes of pleading, Scranton manages to smooth it over. He promises. First thing in the morning. Without fail.

He hangs up and flips on the TV, but the news is over. It's the opening act of NBC's "Saturday Night Live." Comedians Bob and Ray are announcing a contest to pick the new capital of Pennsylvania.

Mortally Wounded.   As the sun rises over the island, there is nothing to indicate that anything has changed.

Under Siege.   A constant flow of men and machinery moves across the north bridge to the island.

High Anxiety. Chairman Joe Hendrie questions NRC staffers at a hearing in the commission headquarters on H Street. Privately, the chairman is becoming more and more concerned about the hydrogen above the core.

Roger Mattson, the NRC's top safety engineer is afraid some of the fuel has melted. He knows there is hydrogen trapped above the core; he suspects it may soon be explosive.

PHOTOGRAPHS BY P. MICHAEL O'SULLIVAN

President Jimmy Carter leaves the island. At almost the same moment, the NRC commissioners in Washington are finally voting to recommend emergency evacuation.

Rosalynn Carter, in a brave show of confidence, talks to the Middletown citizens who stayed behind.

The Snake Pit. Harold Denton (back to camera) waits to be introduced to the press. PR-man Joe Fouchard has created an instant press center in the Middletown Borough Hall gymnasium.

The President's Deputy, Harold Denton, fields questions from a hostile press corps in the Middletown Borough Hall. Joe Fouchard, his weary press aide, stands to his left; staff expert Roger Mattson is behind him.

17

Near the border on Interstate 81, the trooper can see the blue Chevy coming on like a rocket. The son of a bitch must be doing 80. Obviously, this guy is up from Maryland and unfamiliar with Pennsylvania justice. He hits the siren and peels out. The Chevy pulls over immediately, and the trooper stops a short distance back. He looks it over carefully. A carload of businessmen. He adjusts his broad hat and approaches.

And then comes a line any speeder would envy being able to use. "We're from the Nuclear Regulatory Commission," says the man. "That's Chairman Joe Hendrie. We're on our way to Three Mile Island."

The stunned trooper backs away. He shouts, "Move out." As the Chevy pulls back onto the highway, the officer dashes to his car and radios ahead, clearing a path through Pennsylvania for the chairman.

In the backseat with Hendrie, Roger Mattson talks about the hydrogen bubble. Since Friday, when he raised the question before the nation's scientific community, he has been bombarded with answers all across the spectrum, and some of the possibilities are truly horrifying. There is little doubt now that an explosion inside the reactor vessel would be catastrophic. And some of the people Mattson has been talking to think the bubble might already be explosive.

Racing down Highway 141, everyone falls silent as they near the plant. They crest the hill and there it is; sprawling, gargantuan structures covering the island as if it were some alien outpost. "Which one is Unit Two?"

"I think it's the one on the left."

Slate gray and silent, it looks benign enough. But Mattson has X-ray vision. In his mind's eye, he can see inside that massive dome to the south, and he can imagine the core, fractured and seething, fuel rods jumbled in disarray, a hundred and fifty tons of radioactive uranium hanging by a thread beneath a thousand cubic feet of hydrogen. It's a good thing that pack of reporters up ahead can't read his mind.

They pull into the Observation Center, and Joe Hendrie does his best to be unobtrusive. They are already late; the last thing they need is to be recognized by the quagmire of newsmen surrounding the building. One of Hendrie's assistants runs to the NRC trailer to locate Harold Denton. He quickly rushes back with the word that Denton has already left for the airport. They jump back into the car and pull out. It is every top bureaucrat's recurring nightmare to be late for a meeting with the president.

The aging hangar at Harrisburg International now belongs to the National Guard, but there are still paint-peeling hints of the old days when this was Olmstead Air Base. In the echoing emptiness, groups of men cluster, quietly awaiting the arrival of the chief executive. Governor Thornburgh is here. And Harold Denton is here, along with Victor Stello, the hulking director of inspection and enforcement, NRC headquarters. Stello came up on the helicopter with Denton on Friday, and over the past two days he has been functioning as Denton's chief of staff for operations.

There is a commotion at the door. Chairman Joe Hendrie has just arrived. Then Stello spots Roger Mattson. The two men head for each other. Stello shouts, "Mattson, you dumb shit!" He is outraged by the loose-lipped speculation that has been coming out of Washington since he left. He tells Mattson that the people in Bethesda are responsible for a near panic last night.

But Mattson is no pansy. Though Stello outweighs him by a hundred pounds, Mattson stands belly-to-belly and calls him a bull-headed son of a bitch. The staff's calculations, says Mattson, are not based on hot air. For the past forty-eight hours, he and his people have been in contact with the best scientific minds in the country, and regardless of what Stello thinks, some of these people are coming up with answers that show the bubble is going to be explosive.

The heat of this argument seems to shock both of the participants. Stello is astonished at Mattson's vehemence, and vice versa. Both men seem to take a step backward, each one wondering whether the other could be right.

For Harold Denton, this is a thrilling moment indeed. He is standing here in this godforsaken airplane hangar, the president of the United States is arriving at any moment, and his two top staff experts are shouting about whether or not the place is going to blow up.

Mattson is afraid that Stello and the people on-site have been too concerned about hydrogen inside the building to worry about the hydrogen bubble inside the reactor vessel. But Stello says he has considered the bubble and nobody on the island thinks it's going to be a problem. He thinks Mattson's experts are full of crap. All these armchair scientists are forgetting that hydrogen under high pressure will prevent the water from breaking apart into hydrogen and oxygen because it will tend to suppress the creation of more hydrogen. Without free oxygen, there can be no explosion.

And then they all hear it, the heavy crackling that any Viet Nam veteran immediately recognizes, the approach of military choppers. Slowly, gracefully, the presidential heli-

copter settles to the ramp with a WHUMP WHUMP
WHUMP that rattles the hangar doors, and Jimmy Carter
emerges with a wave of the hand. With him, in a brave show
of confidence, is the First Lady.

Mattson is amazed at the entourage that seems to move
with the president like a school of sharks: Secret Service
agents, the White House press corps, the guy with the "football" just in case the Russians are coming. Carter is ushered
in the main door of the hangar as the press is herded in back.
Thornburgh, Denton, and the president pose for photographs, then they kick everybody out and get down to
business.

As the security people stand about the hangar, the president and his aides cluster around Harold Denton to get the
update. Denton has had a cutaway model of the plant
brought here from the Visitors Center. First he gives everybody an overview of the situation. With flip charts and
drawings, he explains the present dilemma. Then he delivers
the punch line.

With a glance toward his two feuding deputies, Denton
tells Carter that there are a couple of schools of thought
among the experts. One school says the hydrogen bubble is
becoming explosive; the other says it is not. But in either
case, he says, there is agreement that nothing will happen for
several days. Denton believes there is no clear and present
danger.

The president has a degree in nuclear engineering, so he
knows the terminology and is able to ask several intelligent
questions. On the subject of hydrogen, he has a suggestion.
He turns to his science advisor, Frank Press. "Frank, tell
these gentlemen what you were telling me on the way up."

Frank Press suggests they try injecting some chemicals
into the coolant system that would combine chemically with
the hydrogen and, in essence, soak it up and render it harm-
less. He suggests they try the element palladium. Mattson

thanks him for the suggestion, but he has already been down this road. It's true that palladium is eager to combine with hydrogen: so eager that it explodes on contact.

As this somber meeting grinds on, the Secret Service checks out the route between the airport and the plant. In the presidential limousine, agents make the run through Middletown to the north gate and back, checking rooftops and windows along the way. It never occurs to them to ask anyone if the president will be safe from radiation. The place is crawling with federal employees; surely these people know what they're doing.

At the north gate, the press has been gathering for the past hour. The network camera crews have already staked out their positions, and one enterprising photographer has climbed up the railroad-crossing signal for a high-angle view. He's lining himself up for a shot of the president with those great ugly cooling towers in the background.

As they wait, the wind slowly begins to haul around to the south. In a matter of moments, the radioactive plume that had been blowing away from them is wafting directly over their heads. Anxious glances are exchanged, but they hold fast. Better cancer than ridicule.

And now comes a black cavalcade heralded by sirens, moving down the river road. Along the way, in the yards of the homes that lie literally in the shadow of the cooling towers, are many of the local people who could not or would not leave. As he passes, they cheer him madly, for they are angry, confused, and afraid, and he is the president of the United States.

Carter, Thornburgh, and Denton, each with a considerable entourage, get out of their limousines and board a school bus for the trip across the bridge. The pool reporters follow, and as the bus moves out, they get a brief lecture on how to behave. Everyone will wear radiation badges and plastic booties. No smoking. Don't touch *anything*.

The scene on the island resembles a siege. Everywhere they look, there are crews of hard hats hauling, welding, hoisting, unloading; and across the bridge grinds an unending line of trucks with loads of steel and lead, tanks, tankers, machinery, and machinists.

The man picked to see Carter through the plant is Shift Supervisor Jim Floyd. It's been quite a week for Floyd. How could he have imagined last Wednesday, when he went down to breakfast at the Sheraton in Lynchburg, that he would be escorting the president up the steps of the control building four days later?

Carter is stunned by his first glimpse of the control room. The scale is overwhelming. He is used to the navy approach, where the whole reactor console is within reach of one or two men. This thing is a gymnasium, an electronic gymnasium with every light in the place flashing a different color. He begins to get an inkling of how it might have all gone wrong.

As he turns to go, the president glances at the back wall. Everyone in the room follows his gaze. He's looking at that poster somebody forgot to take down, a picture of a quivering infant that reads:

"I DON'T KNOW WHETHER TO CRY MY EYES OUT, SCREAM MY HEAD OFF, OR WET MY PANTS."

By now almost everybody in the country has heard about the bubble at Three Mile Island, but one group of souls almost within sight of the reactor remains in the dark. The Amish, who eschew everything having to do with nuclear power including the electricity itself, have no radios or televisions. Only a few miles down the valley, horse drawn, they pursue the old ways, oblivious of the twentieth-century nightmare just over the horizon.

A couple of hundred miles to the south is another band of citizens who, in their own way, seem to be as out of touch with reality as the Amish are. The four NRC commissioners Joe Hendrie left behind have spent the morning with one of their major preoccupations: struggling over the wording of a memo.

"Could I make a slight revision of title?"

"Yeah."

"Call it 'Basis for Evacuation Planning.' "

"I don't know. I think we've got to get straight on what we're doing, so let's be clear on that."

"I tend to view it as the procedures which NRC is involved with in making their evacuation decision. Okay?"

"All right. Okay."

"Now let's start out. Number one, who's going to decide?"

"I guess it's the decision chain, because . . . who's going to decide depends upon which series of events."

"That's—this is what we were talking about before."

"Well, but let's say 'who decides' is the question. And that depends. Okay?"

"Okay, well, let's put that down. Okay. If there is not time —'now no time'—we've got to start talking in terms of——"

"What does 'no time' mean?"

"It's the situation becoming critical suddenly."

"—don't want to use that word."

They are interrupted by Robert Budnitz, deputy director of research for the agency. Budnitz is one of the hydrogen-bubble experts Roger Mattson turned loose yesterday, and for the past twenty-four hours he has been exploring the arcane avenues of speculative engineering.

"We now have two different groups doing calculations," he says, "one in Pittsburgh, one in National Reactor Research Lab, and a third group in Idaho working with them. And we now understand what the flammability problems are with that stuff in the upper head, and I'll give you the numbers, if I can find them." Riffling through his notes, he tells them the best calculations now indicate that if the oxygen level reaches 4.8 percent, the mixture will be flammable.

"Now at 5.0, if you light a spark it will burn . . . When that burn takes place in a ten- or twenty-millisecond process, you get a pressure pulse . . . That's 5,500 pounds per square inch . . . So if you get that burnup there and it goes to 5,500 psi, we're in trouble, and the reason why we're in trouble is with the yield stress of that vessel."

"Well, what does it do?" asks Gilinsky. "Is it a spike?"

"It will cause—it's a hoop-stress problem and it will cause a crack which, if you were looking out, it would be like this." He sketches a line on the diagram.

"What's the detonation——"

"No, the explosion's a little later," says Budnitz. "I'll come to that. This is not an explosion. This is a chemical burn, but it's fast . . . And we might lose that vessel, which we can't afford."

"Do you expect any kind of time sequence?" asks Bradford.

"There is going to be a propagated pulse everywhere in the system. We're going to lose valves; we're going to lose seals; we're going to lose the pumps. We just can't stand that."

"Tell them about the ringing," says John Ahearne.

"The pulse doesn't just happen once and it's over. It's like a wave which will bounce back and forth because there is no way to damp that energy rapidly," says Budnitz. The vision of this destructive shock rebounding through the pipes is enough to turn everyone white. But there is more.

As the oxygen level increases, less energy is required to trigger an explosion. "It might even be induced by things like sloshing," he says. Also there is a possibility that the gas might come in contact with a hot spot inside the core. "As soon as that gas gets to 680 degrees Fahrenheit it will spontaneously combust without any ignition, and that's true at any mixture above the 5 percent . . . At 10 percent we reach a regime where if it starts, it doesn't burn, it explodes. It turns out that you get a pressure pulse which is . . . 13,000 psi. Gonna lose everything if we get that."

And there is one other thing. Budnitz says it's possible that in the complex structure of the upper reactor head, among the steel plates that hold the fuel assembly in place, there are pockets of gas that are already explosive.

"We don't have any idea what the geometry is there," he admits, "and it's been studied carefully: people have looked at drawings. And everything looks like it's flat. But anybody who has ever seen something with bolts on it knows that not everything is flat that's got bolts on it."

In other words, there is a chance it could explode at any minute.

Somebody says they should check on the state of Pennsylvania to see how quickly they could get an evacuation underway. Yesterday, when Victor Gilinsky tried to ask that same question, he was assured that the state could evacuate out to five miles in a couple of hours. But the staff in Bethesda is beginning to discover it's more complicated than that. It's not just a question of grabbing the bullhorn and telling everybody to hit the road. There are prisoners, premature infants in incubators, patients on life-support systems, nursing homes, orphanages, and thousands of people who don't have their own cars; there are invalids and elderly who cannot move without assistance.

Doc Collins, the agency's evacuation expert, has been in touch with Colonel Oran Henderson at the Pennsylvania Emergency Management Agency, and he has some critical information. Not only is it impossible to expect a meaningful evacuation in two hours; Colonel Henderson says he would like four hours' advance warning *before the evacuation begins.*

It will be dark in four hours. At the very least, says Peter

Bradford, they should let Governor Thornburgh in on this new information. A discussion breaks out. Commissioner Kennedy tries to get everyone back to work on the memo, but they can't concentrate. Finally John Ahearne says, "I believe the question we have to try to address is whether or not we think we have to say that we ought to order a precautionary evacuation."

For the past ninety-eight hours, they have managed to steer clear of this decision. On those occasions when it seemed close, Commissioner Kennedy, the War College graduate, rallied them with his determination: Don't Give Up the Ship. But he can hold them no longer; they are breaking out the life boats.

Bradford: Aye.

Gilinsky: Aye.

Ahearne: Aye.

Kennedy still isn't sure. Chairman Hendrie will be meeting with the engineers at the site shortly; Kennedy thinks they should wait and see. Ahearne says, "Okay, can I—is it correct that you're saying that at this point, if there is no new or positive information available after that meeting, you would go along . . ."

"I think so."

It has taken four days to call for a vote on the previous question; when the ballots are finally cast and the tentative decision is made to recommend evacuation, the president of the United States is only a stone's throw from the reactor.

The citizens of Dauphin County, however, are not so naive. Calmly, rationally, without the slightest ripple of disorder, they have already left. Common sense may be in short supply at the NRC, but not in Dauphin County; unlike the officials in Washington, these people decided to play it safe. Thornburgh's office estimates that in the past forty-eight hours, over a hundred thousand of his constituents have headed for high ground. When a reporter pulls in to a station off the turnpike to ask directions to Middletown, the man at the pump says, "Go to the end of this road and turn left at the light. It'll be the second empty town you come to."

Word of the trapped hydrogen has spread ripples far beyond Pennsylvania. Even in Washington, businessmen and minor officials have developed a sudden urge to travel. Personal friends of the commissioners are calling to see if it's time to head south. At the crest of the panic, Lieutenant Governor Bill Scranton gets a call from an old gentleman up in Connecticut. "I'm seventy years old," says the man. "I'm retired, I don't have any friends, I'm alone in the world. If you need somebody to go in there, send me."

But this noble gesture, even if it were possible to carry out, is no longer useful; the people on the island have more than enough courage to go around. At considerable risk, one of the repair crews has managed to run a hose from the overpressured waste tanks in the auxiliary building to a fitting on the side of the reactor building. Now the operators can bleed hydrogen out of the coolant as it passes through the auxiliary building and send the deadly gas back into the containment instead of up the plant stack.

A wall of lead has been built around the hydrogen recombiners—devices hooked to the containment building that slowly burn the hydrogen—and these machines are working. Little by little, the bubble above the core is being bled away. It was Victor Stello's experts who were right, not Roger Mattson's.

18

Monday, April 2

The crackle of anxiety Peter Bradford noticed on his last trip out to staff headquarters has subsided. It was a close call —they are just beginning to find out how close—but the gut-wrenching apprehension is over; he can sense it just walking down the hall. As he turns the corner, he sees several of the agency's top engineers gathered in one the offices, clustered like students around a middle-aged professor in a gray suit and rimless glasses who seems to be explaining exactly how the accident happened. Bradford steps in. "Who's that?" he asks.

"An engineer from the Tennessee Valley Authority. Name is Carl Michelson. Says he warned Babcock & Wilcox about this two years ago."

As the impact of the accident begins to sink in across the country, the scientists—the theorists and pure physicists— are astonished at all this untidiness. In their world, a law is either true or it's not. There is none of this hit-or-miss business that works only when everybody's feeling okay.

Dr. Edward Teller, father of the H-bomb and longtime proponent of nuclear power, is righteously indignant. He and his colleagues privately rail at the stupidity of the operators: had they kept their hands to themselves, the automatic safety systems would have saved the plant. For the engineers, however, it is a learning experience. They have discovered once again that the laws of Newton and Einstein are no more immutable than the law of Murphy: If something can go wrong, it will.

Roger Mattson is an engineer, and he understands the inexact nature of his profession. "An engineer is someone who measures something with a micrometer, marks it with a piece of chalk, and cuts it with an ax." Right now Mattson and the NRC technical staff are all under tremendous pressure to get the reactor out of its present delicate condition and into some kind of final cool-down phase. Even though the situation is more or less stable, the core is riddled with hot spots, the coolant temperature is still over 280 degrees, and the pressure is above 1,000 pounds per square inch. Nobody who knows anything about light-water reactors is going to sleep very soundly until they can find some way to take the damn thing to cold shutdown.

At the highest levels of the NRC, there is great anxiety about the situation. The core is being cooled by a single pump. It could go at any second, and there is no guarantee they could get another one started. Harold Denton, for one, would like to get this place turned off as quickly as possible. But Mattson and Vic Stello refuse to be rushed. "It is a very badly damaged core," says Mattson, "and some analyses indicate that natural circulation won't work. This is a situation that we didn't know of, we didn't comprehend, we didn't analyze; and we're making science as we go along, folks. So just stay where you're at. You're in good shape. Nothing is gonna happen tonight. Let us work on the problem."

But the vulnerability of the island to acts of God is illustrated a few days later. John MacMillan, head man of the Babcock & Wilcox nuclear division, is holding a late-night conference in the company trailer across from the island. MacMillan is talking with half a dozen key aides, and one of the topics on everyone's mind is the possibility of a power failure somewhere else that would shut off electricity to the island. Just then, the lights go out. In a slapstick scramble, they tumble from the darkened trailer and run to see if the lights are still on over on the island. They are. It was only a fuse in the trailer. It might have been funny if they hadn't all been scared shitless.

Then on April 7 comes the jolt they've been dreading. It is shortly after noon and the control room is filled with nuclear experts from all over the country—an ad hoc "industry advisory group"—when the alarm sounds and reactor coolant pump 2A, now the sole source of cooling water for the core, trips off the line and shuts down. The operators leap to the console; they've been rehearsing for this; it was only a question of time. They race through the start-up sequence. A hundred and twenty seconds later, reactor pump 2A is in operation and up to speed.

Although it was anticipated, this heart-thumping little episode seems to fuel Denton's argument: get to cold shutdown as quickly as possible. But over the next several hours, the experts watching in the packed control room can see that the temperature pattern in the core is slowly changing. The core is *moving*.

After this incident, the only people still pushing for a quick solution are the politicians. The engineers have stopped arguing with Mattson; everyone on the inside now realizes that the slightest jiggle could "rearrange the fuel" in a most unpredictable way.

Over the next three weeks, Mattson and Stello lead a vast national effort to chart a safe course away from the rocks.

At their disposal, by presidential request, is practically every scientist and steamfitter in the country who knows anything at all about reactors. Several layers of contingency plans are developed. Detailed procedures are written for every step of the operation, and an elite corps of operators is trained for each contingency. On the island, round-the-clock shifts of construction crews beef up the hardware, adding back-up systems in preparation for the big moment.

When the countdown begins on April 27, the control room at Three Mile Island is manned by a hand-picked crew that includes practically every major figure in the nuclear industry. Heading the list is the venerable Dr. Sol Levy from General Electric. For Roger Mattson, listening in from the Incident Response Center in Bethesda, the scene conjures up a tense image of Stagg Field and Enrico Fermi and the first chain reaction. Never did Mattson imagine he would someday be doing original research on a live power reactor.

At 2:03 P.M., reactor coolant pump 2A is stopped and natural circulation is established for the first time in both boilers. As the temperature readings rattle in to NRC headquarters, teams of analysts compare the heat-transfer plot with their predictions. As the data develop, Mattson and his crew can see the fuel redistribute itself. But it stays within limits, and by nightfall, the once mighty core of Unit Two is on its way to cold shutdown. All that remains is the problem of what to do with this incredible mess.

Jim Creswell predicted it. He felt it in his bones. All the same, it's a little eerie to be standing here looking across the Susquehanna at the absolute proof. Creswell is waiting for the bus to take him across to the island; he wants to get a look at the control room. Then he will begin interviewing the plant operators. This time Creswell will be a part of the investigation. After the press discovered him, he was too hot for the Chicago office to handle, so the NRC assigned him to the Three Mile Island inquiry.

The NRC internal investigation is one of five in progress. Both the House and Senate have hunting parties in the field, the president has named a blue-ribbon commission to be headed by Dartmouth president John Kemeny, and a group of independent consultants has been set up under attorney Mitchell Rogovin. Each of these commissions will have its own investigators, but they will all depend to a great extent on these initial interviews. Creswell can't wait to start digging.

Like everyone else, he is impressed with the scale of the control room. It is big. Right now, that's just as well, because it's full to the gunwales. Someone in the room recognizes him and there is a murmur. A gunfighter steps out of the crowd. A tall, lanky man with a drooping mustache approaches and sticks out his hand. It is Ed Frederick, the senior control-room operator, the man who was running the plant the night of the accident. He looks at Creswell. "You are the guy that raised the issue of pressurizer level in the control room, right?"

"Yeah . . ."

"Nice going."

Frederick turns to leave and Creswell, ever the skeptic, stares after him. How did he mean that?

What was vindication for Jim Creswell was a knife in the dark for Walter Creitz. He was relieved of his responsibilities as president and given the title of "adviser." And GPU, the parent company of Met Ed, which used to measure its profits in the tens of millions of dollars, experienced a net operating loss of $16 million in 1981. The accident, however, is only beginning. The costs of the cleanup are so staggering that bankruptcy is not a practical solution; only a massive infusion of cash from the taxpayers will keep the doors open and keep the existing plants running.

The full dimensions of the problem—the loss of income and the cost of the cleanup—come to light in a strange way. General Public Utilities files a lawsuit against the Nuclear Regulatory Commission for "faulty regulation." Specifically, they allege that the NRC failed to inform them of significant events like the blowdown at Davis-Besse. The suit seeks damages in the amount of 4.3 billion dollars, roughly forty dollars per taxpayer.

At the Versailles Room of the Bethesda Holiday Inn, the Nuclear Regulatory Commission is holding the annual awards banquet. Lee Gossick, the agency's executive director, is a former air-force general, and he has arranged for a military band to play "The Stars and Stripes Forever." Now he rises to the podium, and as he begins reading the citation, the room falls silent.

> ". . . for his skill and tenacity in identifying and bringing to the attention of the NRC management and licensing boards the problems associated with the response of B & W plants to react to transients. He persisted in his efforts to direct the licensee's attention to the problems and to bring these problems to the attention of NRC decision makers. In this way Mr. Creswell set an excellent example of a type of performance that is required for the agency to successfully carry out its safety-regulation responsibilities."

And with that, he hands the astonished Creswell a check for four thousand dollars. There is more than one way to skin a cat.

The NRC has decided to make Creswell and Carl Michelson into showcase examples of the agency's new look. The

NRC has taken so much heat about the way they stifled the warnings from these two individuals that they have established a special department to look into other areas of potential trouble. The department will be headed by Carl Michelson. Working under him will be Jim Creswell. It appears, at last, that the town has elected a decent sheriff.

But the old problems persist. Creswell finds that he is as powerless as ever; the product of his labor continues to be paperwork. In January of 1981, he resigns from the NRC and leaves the industry.

Roger Mattson sits in a cluttered office in the East-West Towers surrounded by stacks of depositions and transcripts, gazing out the window at the rain clouds over Bethesda. The nightmare edge of terror is missing, but he can still remember standing at the precipice. He is a man changed in some fundamental way. "There is a difference between believing that accidents will happen and not believing it," he says. "The fundamental policy decision made in the mid 1970s—that what we had done was good enough, and that the goal of regulation ought to be to control the stability of the licensing process—that the body of requirements was a stable and sufficient expression of 'no undue risk' to public health and safety—and if new ideas come along we must think of them as refinements on an otherwise mature technology—that's been the policy undertone. That's the way I've been conducting myself . . . It was a mistake. It was wrong. There were holes in that . . . state of maturity. Mostly having to do with human beings."

One of the critical failures uncovered by the investigators was the lack of direct communication between the control room and the NRC. The insurance company was able to get through to Unit Two almost the minute Bill Zewe declared a site emergency, but it took three days and a presidential order before a hard wire was set up between the plant and NRC headquarters.

A permanent fix was designed to prevent anything like this from ever happening again. Shortly after the accident, every nuclear-power-plant control room in the country had a hot-line telephone installed that connects directly to Bethesda by just a picking up of the receiver. The Incident Response Center is now manned round the clock. At night, the duty officer carries a "potty phone" radio link that keeps him in touch even if he is momentarily out of the room.

But in the log book tonight is a warning that a dozen of the new lines are not working. On the console, one of the lights is flashing; the duty officer says it's a false alarm.

Bob Pollard was once a licensing project manager for the NRC. Like Jim Creswell, he resigned because he couldn't get the agency to do anything about safety violations. Pollard was assigned to the Indian Point Reactor, a Consolidated Edison plant thirty minutes upriver from Manhattan.

When he quit, he went to work for the Union of Concerned Scientists. Throughout the accident, he was a key technical source for the press. Now that the dust has settled, several reporters have gathered in his cramped Washington office to see if they can get a fix on the bottom line: what did it all mean? Somebody asks if the whole thing might not have been blown out of proportion. "According to the reports, relatively little radioactive fallout actually escaped. Nobody died. And in the end, the safety systems worked. At least the containment worked. Was it as serious as we made it out to be?"

"You don't have the ultimate catastrophe out of the blue," says Pollard. "There's always some kind of warning. This was a warning."

GLOSSARY OF TERMS AND MAJOR FIGURES

AEC (Atomic Energy Commission). Predecessor to the NRC, the AEC oversaw the creation of the A-bomb and managed the civilian nuclear-power program until the Seventies.

Abraham, Karl. Public-affairs officer for NRC Region One; drove to Harrisburg and became the liaison officer between the NRC and the governor's office.

Ahearne, John F. One of five coequal and independent commissioners who run the NRC; appointed by the president.

Annunciator. A warning light and audible signal indicating that something in the system is changing or malfunctioning.

Auxiliary building. Essentially an enormous pump house, this building houses all the reactor support equipment (compressors, pumps, tanks) that does not have to be inside the containment building.

Auxiliary feedwater pumps (emergency feedwater pumps). Three independent high-pressure pumps located in the auxiliary building that can take water from several sources and pour it into the reactor coolant system. In certain situations (e.g., a sudden pressure drop) they come on automatically. They are part of the *emergency core-cooling system.**

B & W (Babcock & Wilcox). Designers and builders of the reactor at Three Mile Island. The nuclear division of this century-old manufacturing company is headquartered in Lynchburg, Virginia.

Blowdown. A sudden loss of pressure in the reactor coolant system.

*Italic entries are also identified in this glossary.

Boiler. Technically known as a *steam generator* in the nuclear industry, the boilers are heated by 600-degree water from the core. Pure water on the outside of the pipes is under lower pressure, so it boils, generating steam.

Bradford, Peter A. One of five coequal and independent commissioners who run the NRC; appointed by the president.

Chain reaction. Radioactive decay in an element like uranium releases neutrons that split other nearby uranium atoms. They in turn release other neutrons; if everything is exactly right, the process is continuous and self-sustaining.

China syndrome. Some scientists feel that if the nuclear fuel in a reactor reaches the melting point—5,200 degrees—the melting process will continue until the ball of molten fuel bores through the bottom of the plant and into the ground.

Containment building. The bullet-shaped, twenty-story concrete-and-steel structure that houses the reactor; at Three Mile Island, the containment is designed to withstand the direct impact of a jetliner.

Control rods. Slender tubes of boron or cadmium that absorb neutrons; by moving them in and out of the core, the operators control the level of the chain reaction and thus the amount of power generated by the reactor. In a SCRAM, the rods are slammed into the core to stop the reaction.

Core. The uranium fuel that powers the reactor. At Three Mile Island, the 150-ton core consists of 36,000 finger-sized rods of *Zircaloy*, each twelve feet long and filled with uranium pellets.

Creitz, Walter. President of Metropolitan Edison.

Creswell, James. Reactor inspector, NRC Region Three, Chicago.

Critchlow, Paul. Press aide to Richard Thornburgh, governor of Pennsylvania.

Curie. A unit of radiation measure equal to 37 billion radioactive disintegrations a second.

Decay heat. Heat produced in the core by the decay of radioactive garbage leftover from the fission process. It continues for all time and must be continuously removed or the core will begin to reheat.

Denton, Harold. Director, Division of Nuclear Reactor Regulation, NRC headquarters.

ECCS (emergency core-cooling system). The collection of high- and low-pressure pumps, flood tanks, and heat exchangers designed to cool the core in the event of a loss-of-coolant accident.

Emergency feedwater pumps. (See *auxiliary feedwater pumps.*)

Emergency Response Center. The accident command post located in the NRC staff headquarters building in Bethesda, Maryland.

Faust, Craig. Met Ed control-room operator.

Fouchard, Joseph. Director of public affairs, NRC headquarters.

Frederick, Edward. Met Ed control-room operator.

Fuel rods. Hollow twelve-foot tubes of zirconium alloy that hold finger-sized pellets of uranium fuel. These tubes are the first line of defense against a leak in the system. If they crack, radioactive gas is released into the cooling water.

Gallina, Charles. Investigation specialist, NRC Region One.

Gilinsky, Victor. One of five coequal and independent commissioners who run the NRC; appointed by the president.

GPU (General Public Utilities). The parent company of Metropolitan Edison; headquartered in New Jersey.

Henderson, Oran. Director, Pennsylvania Emergency Management Agency.

Hendrie, Dr. Joseph M. Chairman of the five-man commission that runs the NRC; appointed by the president.

I & E (Inspection & Enforcement). The NRC division responsible for seeing that the nuclear-power-plant operators stick to the rules; director, Victor Stello.

Incident Response Center. (See *Emergency Response Center.*)

Kennedy, Richard T. One of five coequal and independent commissioners who run the NRC; appointed by the president.

Keppler, James. Director, Division of Inspection and Enforcement, NRC Region Three, Chicago.

Kunder, George. Met Ed supervisor of technical support, Unit Two.

LOCA (loss-of-coolant accident). An accidental leak in the reactor coolant system; most feared is the major pipe break that might result from an explosion or an earthquake, but this can also result from a stuck-open valve.

MacMillan, John. Babcock & Wilcox vice-president in charge of the nuclear division.

Makeup System (makeup pumps; makeup tanks). A collection of pumps and reserve tanks located in the *auxiliary building* which supplies water to the reactor coolant system as needed.

Mattson, Roger. Director, Division of Systems Safety, nuclear-reactor-regulation branch, NRC headquarters.

Mehler, Brian. Met Ed shift supervisor, Three Mile Island.

Meltdown.   The melting of fuel in a nuclear reactor following a loss of coolant to the core.

Met Ed (Metropolitan Edison).   A subsidiary of General Public Utilities; owners and operators of Three Mile Island nuclear station.

Michelson, Carlyle.   Systems engineer, Tennessee Valley Authority; consultant to the NRC.

Miller, Gary.   Met Ed station manager for Three Mile Island.

Millirem.   One one-thousandth of a *rem.*

NRC (Nuclear Regulatory Commission).   The federal agency created to regulate the design, construction, and operation of nuclear-power plants; originally part of the Atomic Energy Commission.

Observation Center (Visitors Center).   A two-story brick building on the riverbank directly east of the power plant at Three Mile Island. Normally a tourist information center, it became the company command post during the accident.

Power relief valve.   Technically known as a power-operated relief valve (PORV) and known as an electromatic operated valve (EMOV) by its manufacturer; it sits on top of the pressurizer and is designed to handle small bumps in system pressure by opening and then, one hopes, reclosing. It failed to reclose on the morning of March 28.

Pressurizer.   A forty-foot-tall tank connected to the reactor coolant system that controls the system pressure and water level. A bubble of air in the top of the tank can be altered in size (sprays make it smaller, heaters make it bigger), thus adjusting the force pushing down on the water.

Reactor.   The nuclear heat source for the power plant, it consists of the core, coolant system, and boilers.

Reactor building.   (See *containment building.*)

Reactor coolant pumps. The four giant pumps that move the reactor cooling water through the core and into the two boilers.

Reactor coolant system. Keeps the core cool in normal operation; consists of four main pumps, main coolant pipes, boilers, and support equipment.

Reactor vessel. Also known as the pressure vessel; a forty-foot-tall, 800-ton, forged-steel capsule that contains the uranium core.

Rem (Roentgen equivalent man). A standard unit of radiation dose; an exposure of no more than three rem is allowed nuclear workers in any ninety-day period, no more than five rem per year; the allowable public exposure is expected to be much less. A *millirem* is one one-thousandth of a rem.

Safety Valves (safeties, code safeties). High-pressure valves set to blow open when the pressure in the system gets above specifications.

SCRAM. A term from the early days of nuclear research; an emergency shutdown of the reactor, usually accomplished by rapidly inserting the control rods into the core to stop the chain reaction.

Scranton, William. Lieutenant governor of the state of Pennsylvania.

Steam Generator. (See *boiler.*)

Steam table. A simple graph that shows the boiling temperature of water over a range of different pressures.

Stello, Victor. Director of inspection and enforcement, NRC headquarters.

Teletector. A hand-carried radiation detector with a telescoping probe that can be extended about twelve feet.

Thornburgh, Richard. Governor of the state of Pennsylvania.

Transient. An engineering euphemism for an abnormal condition.

Trip. A sudden shutdown of a pump or other equipment; usually automatic, can be manual; similar to a fuse blowing.

Turbine. Propelled by high-pressure steam, the turbine's block-long spinning shaft turns the electric generator.

TVA (Tennesee Valley Authority). A utility company created by the federal government, the TVA manages dozens of power dams and several coal and nuclear plants in the southern states.

Uranium. An unstable (radioactive) element used as fuel in nuclear-power plants.

Visitors Center. (See *Observation Center.*)

Zewe, William. Met Ed shift supervisor, Three Mile Island.

Zircaloy. An alloy of the element zirconium used to make the fuel rods in a nuclear core.

INDEX